THE MAKER'S
GUIDE TO THE
ZOMBIE
APOCALYPSE

THE MAKER'S GUIDE TO THE ZOMBIE APOCALYPSE

DEFEND YOUR BASE WITH SIMPLE CIRCUITS, ARDUINO, AND RASPBERRY PI

SIMON MONK

NO STARCH PRESS
SAN FRANCISCO

Printed in USA

First printing

19 18 17 16 15   1 2 3 4 5 6 7 8 9

SUSTAINABLE FORESTRY INITIATIVE   Certified Sourcing
www.sfiprogram.org
SFI-00854

Text stock is SFI certified

ISBN-10: 1-59327-667-2
ISBN-13: 978-1-59327-667-6

Publisher: William Pollock
Production Editor: Serena Yang
Cover and Interior Design: Beth Middleworth
Illustrator: Miran Lipovača
Developmental Editor: Jennifer Griffith-Delgado
Copyeditor: Paula L. Fleming
Compositor: Serena Yang
Proofreader: James Fraleigh
Indexer: BIM Indexing & Proofreading Services

For information on distribution, translations, or bulk sales, please contact No Starch Press, Inc. directly:

No Starch Press, Inc.
245 8th Street, San Francisco, CA 94103
phone: 415.863.9900; info@nostarch.com
www.nostarch.com

Library of Congress Cataloging-in-Publication Data
Monk, Simon, author.
   The maker's guide to the zombie apocalypse : defend your base with simple circuits, Arduino, and Raspberry Pi / by Simon
Monk.
        pages cm
   Includes index.
   Summary: "A collection of DIY hardware projects using circuits, Arduino, and Raspberry Pi to store electricity, detect
invading zombies, generate solar power, and create communication and surveillance devices. Projects include alarms,
low-power LED lighting, an FM radio frequency hopper, a periscope, a wind turbine, and flash, movement, and noise makers"--
Provided by publisher.
   ISBN 978-1-59327-667-6 -- ISBN 1-59327-667-2
 1. Electronic apparatus and appliances--Design and construction--Amateurs' manuals. 2. Microcontrollers--Amateurs' manuals. 3. Electronic circuits--Amateurs' manuals. 4. Arduino (Programmable controller)--Amateurs' manuals. 5. Raspberry Pi
(Computer)--Amateurs' manuals. I. Title.
   TK9965.M673 2015
   621.381--dc23
                              2015023925

To Jonathan and Michaela
on your wedding.
May you have a wonderful
and zombie-free life together.

ABOUT THE AUTHOR

Simon Monk is a full-time author and maker, mostly writing about electronics for makers. Some of his better-known books include *Programming Arduino: Getting Started with Sketches*, *Raspberry Pi Cookbook*, and *Hacking Electronics*. He is also the co-author of *Practical Electronics for Inventors* and wrote *Minecraft Mastery* with his son, Matthew Monk.

Simon also writes for *MagPi* magazine and helps out with Monk Makes (*http://www.monkmakes.com/*), a company run by his wife Linda, which makes and sells component kits and other products related to Simon's books. You can follow Simon on Twitter where he is @simonmonk2 and find out more about his books at *http://www.simonmonk.org/*.

ABOUT THE TECHNICAL REVIEWER

Jeremy Blum is a "Hardware Astronaut" at Google, where he focuses on electrical design and advanced prototyping for future Google hardware. Jeremy received both a bachelor's degree and a master's degree in electrical and computer engineering from Cornell University, and was selected by the American Institute of Electrical and Electronics Engineers as the 2012 New Face of Engineering.

Jeremy's popular Arduino tutorial videos and his book, *Exploring Arduino*, have introduced millions of people around the world to engineering. He offers engineering consulting services through his firm, Blum Idea Labs LLC, and he frequently teaches engineering courses to young students and adults across the United States. Jeremy's passion is improving people's lives and our planet through creative engineering solutions; you can learn more about him and his work at *http://www.jeremyblum.com/*.

BRIEF CONTENTS

CONTENTS IN DETAIL

ACKNOWLEDGMENTS

Many thanks to the enthusiastic and dedicated team at No Starch Press, especially to my thorough and patient editors Jennifer Griffith-Delgado and Serena Yang, who have guided this project from initial idea to finished book with skill and imagination.

I would especially like to thank Miran Lipovača for his wonderful illustrations that add so much to the book and Jeremy Blum for his technical review of the material. I am very honored to have two such noteworthy individuals involved in the project.

Finally I would like to thank Linda (see Figure 5-15) for her patience and understanding during the writing of this book.

INTRODUCTION

This is a book for people who like to make things but also enjoy the premise of a postapocalypse world where you cannot assume a limitless supply of electricity and other resources. As such, book starts with projects for generating electricity through solar and pedal-power (using a scavenged car alternator). Once you have power, you'll move on to surveillance and monitoring projects that will help you protect your base. Finally, you'll build communication projects that allow you to find other survivors and even send messages to members of your group via silent haptic communication.

KEY MAKER SURVIVAL SKILLS

Some projects in this book require no more technical skill than being able to attach a wire to a screw terminal, while others require you to be able to solder. The detailed primer in Appendix B will get you started on any technical skills that may be new to you. A few of the projects also require some basic wood-working skills and tools, but you won't need anything more fancy than a saw, drill, and chisel.

The more technical projects in the book make use of the popular Raspberry Pi and Arduino as low power easy to use control modules. See Chapter 5 for some Raspberry Pi basics and Appendix C for a primer on the Arduino.

ABOUT THE APOCALYPSE SURVIVAL PROJECTS

Although these projects are all things that are intended to be useful post-apocalypse, most are also useful even if the zombies don't take over. Many of the Arduino projects can actually be grouped together onto a single Arduino, with a single program integrating their different functions.

Chapter 1: Apocalypse Basics gives an overview of what the world will be like when the zombie apocalypse strikes so you know what you're up against. From there, we dive straight into the projects.

Chapter 2: Generating Electricity has two projects. "Project 1: Solar Recharging" shows you how to charge a car battery using solar power, and "Project 2: Bicycle Generator" describes how to use a scavenged car alternator and pedal cycle to charge a car battery.

Chapter 3: Using Electricity covers two projects to get you acquainted with using those car batteries and an Arduino. First, "Project 3: LED Lighting" has you string up some LED lighting to illuminate your base from 12V batteries. Then, you'll turn an Arduino into "Project 4: Battery Monitor" to make sure you don't run out of juice.

Chapter 4: Zombie Alarms includes two builds you can install around your base to alert you when zombies are about. "Project 5: Trip Wire Alarm" is a nice low tech alarm that uses a microswitch and car horn, and "Project 6: PIR Zombie Detector" is a more high-tech zombie detector that uses a motion sensor.

Chapter 5: Surveillance and Raspberry Pi equips you to monitor your undead neighbors from a distance. "Project 7: Monitor Zombies with a USB Webcam" shows you how to set up a webcam and analyze the video for movement detection in Python. "Project 8: A Wireless Zombie Surveillance System" helps you make your surveillance system more practical by extending your observing range with a low-cost Wi-Fi webcam.

Chapter 6: Add Remote Access and Detect Open Doors helps you keep the zombies out of your base. "Project 9: Remote Door Lock" shows you how to set up a door with an electro-mechanical door latch so that you can unlock it remotely using a wireless option, and "Project 10: Door Sensor" will notify you when someone (or something) opens your door.

Chapter 7: Environmental Monitoring contains projects to protect you against other postapocalyptic hazards, because zombies aren't the only things you have to worry about when you're trying to survive. "Project 11: Quiet Fire Alarm" shows you how to hack a loud smoke detector into a more discreet alarm that's integrated with Arduino. Then, you can use "Project 12: Temperature Alarm" to monitor temperatures and set alarms to avoid burst pipes or other disasters.

Chapter 8: Building a Control Center for Your Base lets you take all the sensor feedback, detection systems, and so on from previous projects and monitor everything on one screen. In "Project 13: A Raspberry Pi Control Center," an Arduino monitors for door and zombie movement, temperature changes, and battery warnings, while the Raspberry Pi displays a status window. You can also add wireless communication between the Raspberry Pi and Arduino on the Control Center by following the instructions for "Project 14: Going Wireless with Bluetooth." This way, you'll put more distance between you and the dangers you're monitoring.

Chapter 9: Zombie Distractors shows you how to draw zombies' attention away from you so you can escape. "Project 15: Arduino Flash Distractor" uses scavenged disposable camera flashes to distract zombies, and "Project 16: Arduino Movement and Sound Distractor" does the same job with the sounder from a smoke alarm and a servo motor waving a flag.

Chapter 10: Communicating with Other Survivors shows you how to find other humans in the zombie-ridden wasteland that used to be your town. Use "Project 17: A Raspberry Pi Radio Transmitter Beacon" to attract fellow survivors with an FM transmitter, hack a low-cost FM radio to scan the air waves for messages from fellow survivors with "Project 18: Arduino FM Radio Frequency Hopper," and flash out messages to would-be recruits to your survivors group (or warn people away) with "Project 19: Arduino Morse Code Beacon."

Chapter 11: Haptic Communication is an essential build if you want to coordinate a group of survivors on a supply run, and it's probably the coolest project in the book. "Project 20: Silent Haptic Communication with Arduino" allow you to press a button on one device and have the other device vibrate (and vice-versa). The project uses an Arduino, low cost 2.4GHz RF modules, and a vibration motor.

Now that you've seen an overview of the projects that will save you from the zombies, you might like to order some parts. Each project includes a parts list specific to that build, including quantities, and Appendix A provides details of where you can buy all the parts listed for each project in the book.

RESOURCES TO DOWNLOAD BEFORE THE ZOMBIES RISE

The book has its own web page at *http://www.nostarch.com/zombies/*, where you will find further information about the book including errata and links to the source code used in the project. That code is all available on GitHub at *https://github.com/simonmonk/zombies/*.

Before the apocalypse strikes, be sure to visit both URLs, download all necessary files for the projects, save them to a flash drive, and keep that flash drive in your go bag. The Internet will very likely cease to exist during the apocalypse, whether because everyone at your ISP becomes a zombie or because the electrical grid itself collapses, but if you download these files ahead of time, you'll be one step closer to outlasting the undead.

With your files loaded and ready, let's look at what you can expect to happen during the apocalypse.

1

APOCALYPSE BASICS

Before you start working on the
zombie apocalypse survival projects
in this book, I want to show you exactly
what kind of undead you'll be dealing with
and share some tips about how to survive in a
zombie-infested world.

Of course, you'll need parts to make your projects. Fortunately, one of
the few benefits of a postapocalyptic world is that there's plenty of scrap
material to scavenge! So in this chapter, I also include a guide to finding the
parts you'll need.

But first, let's look at the background of zombies.

ZOMBIES

I find that people tend to consider themselves either zombie lovers or zombie-indifferent. Since you're reading this book, there is a good chance you're a zombie lover like me.

The appeal of zombies lies both with the zombies themselves and in the postapocalyptic scenario that the survivors face. You could likely defeat a single, slow zombie quite easily: a baseball bat to the head should do the trick nicely. But in numbers, zombies become a serious threat.

If you look up "Zombie" in Wikipedia, you'll find two entries: "Zombie (fictional)" and, rather worryingly, just "Zombie." The nonfictional zombie is, according to Haitian folklore, a corpse that can be raised by magic to do its master's bidding. These folklore zombies are never going to be present in significant numbers to cause the sort of apocalypse portrayed in popular culture. For a situation where most of the human race has died or been *turned* into a zombie, we need some *fictional zombies*.

TYPES OF ZOMBIES

Fictional zombies have roots in 19th-century fiction, with Mary Shelley's *Frankenstein*, but they became prominent in modern times through movies such as *Night of the Living Dead* (Figure 1-1).

FIGURE 1–1: ZOMBIES FROM Night of the Living Dead

The zombies depicted in *Night of the Living Dead* are classic *slow zombies*. Slow zombies shuffle around as if in a daze, searching for human flesh to eat. Interestingly, the zombies in this movie are able to use tools, breaking windows with rocks and bashing doors with hefty sticks. Most zombies lose this skill in later film and TV depictions. The portrayal of slow zombies may have reached its cultural zenith with the hit TV show *The Walking Dead*.

Slow zombies are the most common fictional zombies, and this book focuses on the threats they pose. There are, however, many other types of zombies, as different filmmakers have sought to put their own imprint on the concept. Table 1-1 lists some of the most important modern zombie portrayals along with some features of each type of zombie.

Table 1–1: FICTIONAL ZOMBIE VARIETIES

FICTIONAL DEPICTION	FAST/ SLOW	EATS	ALIVE/DEAD	CAUSE OF OUTBREAK	MEANS OF DISPATCH
Night of the Living Dead	Slow	Human flesh	Dead (reanimated)	Radiation	Head trauma
Hell of the Living Dead	Slow	Human flesh, other zombies	Dead (reanimated)	Chemical leak	Head trauma
Return of the Living Dead	Slow	Human flesh, especially brains	Dead (reanimated)	Chemical leak	Head trauma
Resident Evil	Slow	Human flesh	Alive	Virus	Head trauma
World War Z	Fast	Human flesh	Alive	Parasite/ virus	Head trauma
28 Days Later	Fast	Human flesh	Alive	Virus	Normal means
Shaun of the Dead	Slow	Human flesh	Dead (reanimated)	Unknown	Head trauma
The Walking Dead	Slow	Human flesh	Dead (reanimated)	Unknown	Head trauma

All zombie types have a number of factors in common. Chief among these is a desire for human flesh. Another, almost universal, truth is that the only sure way to kill a zombie is severe head trauma. Decapitation is highly effective.

ARE ZOMBIES REALLY DEAD?

One important question is whether a person has to be dead in order to qualify as a zombie. In some films, such as *World War Z*, the zombies are not dead but rather living humans who have been mentally altered by a virus or other parasite. Some would argue that such creatures are, strictly speaking, not zombies at all.

The terminology of death is also tricky with zombies. If a zombie is already dead, how can you kill it again? Although the zombie is a person who has died, has the process of zombification actually brought the person back to life? In that case, zombies could certainly be killed for a second time.

However, we often define death as occurring when the heart stops, and zombies' circulatory systems are clearly not functional, as illustrated by their relative immunity to being shot anywhere but in the head.

If zombies are still dead, then it seems wrong to speak of *killing* them, but until popular culture invents a new word, it'll have to do. In this book, I am going to use the phrase *killing zombies*—while possibly inaccurate, it is not ambiguous.

HOW LONG WILL THE ZOMBIES BE AROUND?

Just how long could a zombie apocalypse last before the zombies disappear? This depends, of course, on the rate at which new zombies are created and the rate of zombie death. The curves for human and zombie populations can be plotted along a horizontal axis that shows the passage of time and a vertical axis that shows the population in billions (Figure 1-2).

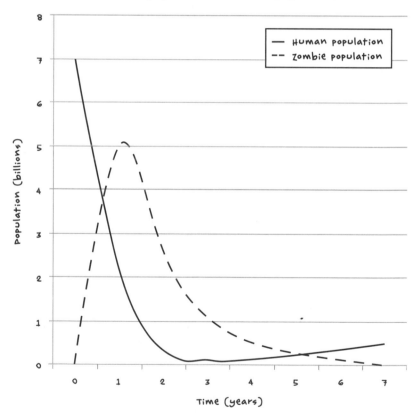

FIGURE 1-2: HUMAN/ZOMBIE POPULATION CURVES OVER TIME

As the outbreak starts, the human population will plummet as the zombie population increases rapidly due to zombification. However, since many humans will be eaten rather than turned, the zombie population will not reach the preapocalypse human level. How much it rises will depend on the zombified-to-eaten ratio, as well as death rates for both zombies and humans.

After reaching a peak, the zombie population will then start to decline. That's because as the human population declines, the surviving humans will be those best equipped to survive. (Perhaps they read this book!) The human population will also become more spread out, making it harder for zombies to find people. Eventually, the population of humans will stabilize at a low level.

Zombies, on the other hand, are unlikely to survive well over the long run. Judging by their hunger for human flesh, they need to eat to survive, although just how this works without a fully functioning digestive system is a mystery. In any case, since they don't photosynthesize, their energy must come from somewhere, and human flesh is the most likely source. But as humans learn to survive, the zombie population will struggle to find food. And since zombies are essentially slow-moving piles of rotting flesh, they're a carrion eater's equivalent of a takeout dinner. If we added population curves to the graph for crows, foxes, wild dogs, and other such animals, we would probably find a massive spike in their populations as they cleaned up the mess in pretty short order. Thus, humans who fight back and plenty of natural predators will put downward pressure on the zombie population.

It also seems extremely unlikely that zombies will breed (something that really doesn't bear thinking about). Therefore, after a while all the zombies will be gone, and humans, who will breed, will start to rebuild civilization.

So, this is your chance. Being in possession of this book should seriously increase your chances of surviving and, hence, reproducing!

POSTAPOCALYPSE SURVIVAL 101

Aside from the emotional tension of the zombie threat, one of the most interesting ideas explored in many zombie films is how the human survivors of a zombie apocalypse will cope.

When the zombie apocalypse comes, you'll need to be prepared. This book will serve as your survival guide.

HOME

Where you live will be critical to your chances of survival. Most suburban homes will not survive an attack from a determined group of zombies intent on chowing down. If you're in such a location, you should probably find a new base pretty early on.

Make sure your new home is easy to defend. Many consider a boat the best place to live (zombies are really bad at swimming!), but this isn't practical if you live a long way from open water. Also, living on a boat presents its own difficulties, including storms and a constant need for fuel unless it's under sail. You will also have to venture onto land to stock up on supplies, but supply runs will be necessary no matter where you live.

A boat is advantageous because it opens up the possibility of finding a zombie-free island where a community could be established. This is definitely an option to work toward. In fact, assuming you survive the first few days, working your way toward the coast or the shore of a large lake in a series of hops is probably a wise strategy. A long drive could, in theory, take you from anywhere in the country to the coast, but in all likelihood, the roads will be jammed with abandoned cars following desperate attempts to avoid the contagion. So, travel is likely to be slow and dangerous, menaced by the ever-present zombies and possibly other survivors.

If you live somewhere where it gets cold in winter, then you might want to consider getting somewhere warmer. Cold weather means that you'll need to consume more calories and find some shelter with heating that isn't too drafty. The only likely way to heat your abode is by burning wood, which you'll have to go out and gather. On the other hand, an ax is an effective weapon for decapitating zombies.

If you know how to fly a light aircraft, then this is a great way to avoid both zombies and ground obstacles. You may find that your destination airstrip is not clear, and many fields will revert to scrubland without human cultivation, so take some exploratory trips before you commit yourself to soaring past the point of no return.

WATER

The survival expert Cody Lundin has something called the Rule of Threes. This can be paraphrased as follows:

- You can live 3 minutes without air.
- You can live 3 hours without shelter (in extreme temperatures).
- You can live 3 days without water.
- You can live 3 weeks without food.

Air shouldn't be a problem and, assuming that the apocalypse takes place during a clement patch of weather, neither should shelter. So apart from avoiding being eaten, your main priority has to be finding potable water and other things to drink. The public water supply is unlikely to keep going if the pumps that pressurize it lose power. Therefore, if possible, find a location with its own well or other freshwater reservoir. Bottled and canned drinks should also be in great supply, as there won't be many people putting their change into vending machines.

FOOD AND FUEL

Farming can take years to become well established enough to feed a small group of people, so growing your own organic veggies is a goal for the future, when it's time to rebuild society. With fewer people around, however, there will be plenty of canned food to find—enough to last almost indefinitely. Scavenge cans of food and other nonperishable edibles from homes and supermarkets.

The projects in this book concentrate on electricity. Not just any electricity but electricity stored in batteries. This is fine for lighting, alarms, and communications, but when it comes to heating and cooking, it is not feasible to use electricity without a serious solar panel array and some heavy-duty equipment. When you're in the mood for hot food then, gas-powered heaters and camp stoves are much more realistic alternatives. Be sure to use them safely!

A barbecue grill is another option for cooking your food and will happily burn charcoal or wood.

ZOMBIE KILLING

By far the best strategy when dealing with zombies is to avoid attracting their attention whenever possible. Try to be quiet and move stealthily as you scout new places, and avoid going anywhere where you might get trapped, including buildings or rooms with only one door as well as blind alleys.

Eventually, you will have to fight a zombie, so make sure that you are always armed. Guns are not necessarily the best option. They make a lot of noise, and they have to be reloaded. Also, to take down a zombie with a bullet, you need to shoot it in the head, and they won't normally stand still while you take aim.

An ax, baseball bat, or sword can be more effective. This was demonstrated in the "Zombie Special" episode of *Mythbusters*, where it was scientifically proven, to a high standard, that you can kill far more zombies per minute with an ax than with a gun. The relative merits of various weapons are listed in Table 1-2.

Table 1-2: WEAPON PROS AND CONS

WEAPON	PROS	CONS
Ax	Excellent for decapitation, inflicting head trauma	May get stuck in thick skulls
Baseball bat/club	Effective for head crushing, with no danger of getting stuck	May require repeated bashing of skull; wooden bats are liable to break
Handgun	Good at close quarters	Noisy, requires reloading
Hunting knife	Good at very close quarters	Requires close contact with zombie, increasing risk of infection
Iron bar	Effective for head crushing; no danger of getting stuck	Heavy
Rifle	Great for long-range protection	Slow at close range
Samurai sword	Very cool! Excellent for decapitation	Potential sticking problems, similar to ax

In fact, different weapons will come into their own in different situations, and ultimately your zombie-killing instrument comes down to personal choice. I favor the often neglected iron bar as my weapon of choice. *Half-Life* players will be well aware of the effectiveness of this weapon.

Regardless of what you pack, fighting zombies is extremely risky. Setting traps to kill zombies from a distance is much better than taking them on at close range. A pit with bait hung over the opening will often be sufficient to induce zombie after zombie to fall down the hole. A mine shaft is ideal for this, as any hole you dig is unlikely to be deep enough to keep the zombies from climbing back out over each other as it fills up.

Thinning out the zombie population in your area will help to reduce the chance of zombies attacking in unmanageable numbers, and it's the socially responsible thing to do for any other survivors that might be in the area. It's the postapocalyptic equivalent of cleaning up after your dog.

When your zombie situation gets more hectic, have a supply of Molotov cocktails (homemade incendiary bombs using bottles and often gasoline) on hand to hurl at zombies from a safe distance. Other projectiles, such as grenades, can also be effective if you can get your hands on them.

DRESSING TO KILL

Whether fighting zombies or trying to escape them, it's important to dress well. That is, no long hair or loose clothing. Once a zombie gets hold of you, it will drag you inexorably toward its mouth until you get into biting range. In other words, wear the kind of clothes you would around machines in a workshop: no long hair to grab and definitely no neckties.

Armor can be improvised. Something as simple as thick cord around your forearms can prevent a bite from penetrating the skin. But don't forget to balance mobility with protection. A medieval suit of armor might provide good protection for a time, but it will slow you down significantly (Figure 1-3). It's hard to put up much of a fight when you can't get away!

FIGURE 1-3: A HEAVY SUIT OF ARMOR CAN DECREASE YOUR MOBILITY.

You should also think carefully about the dangers of becoming infected from blood spatter when killing zombies. Try to breathe through your nose while fighting or even wear a face mask.

STAYING HEALTHY

Healthcare in a postapocalyptic world is a pretty do-it-yourself affair. You'll be lucky to find a medic in your group, as medical staff will have been very much in the thick of it during the initial outbreak and therefore quite unlikely to have survived.

This means that you need to keep yourself healthy. Getting enough exercise is not going to be a problem. Just staying alive, without all the modern conveniences that we take for granted, takes quite a lot of effort. However, you need to stay well to survive, and that includes paying special attention to any minor injuries. All cuts and open wounds must immediately be dressed with antiseptic and covered with a bandage or dressing. You should also keep a stock of antibiotics. If you can't get your physician to prescribe them now, then raiding a hospital or pharmacy will be a priority once the zombie apocalypse starts.

Boil any drinking water not from a sealed bottle, and don't eat anything likely to give you food poisoning.

If you are nearsighted, then a spare pair of glasses is an essential item. Not being able to see well could easily be fatal in this new world.

BE PREPARED

Boy Scouts and Girl Scouts probably already have a special badge for zombie fighting; if not, they will after the apocalypse! In any case, their motto is a good one: *be prepared*. Always think ahead, adopting the astronaut mentality of anticipating the next thing that could kill you as well as the thing after that (if you have time). Rehearse scenarios in your head continually to minimize the chances of disaster when something takes you by surprise.

Keep a *go bag*. This small backpack should be kept close at all times so that you can grab it and run at a moment's notice. It should contain just enough to keep you alive for a few days. A backpack leaves your hands free for fighting. A good contents list for the backpack might look like this:

- Bottled water
- High-energy food such as chocolate and cookies
- Multipurpose pocket knife
- Thermal blanket
- Flashlight
- Spare weapon

Wherever you are, always make sure that there is more than one way to get away. You need a front door and a back door. No matter how impenetrable you think your base is, there is always the possibility that the worst will happen, so have an escape route.

OTHER SURVIVORS

Teaming up with other survivors can be a mixed blessing. On the one hand, the more people, the more food and drink you need. On the other hand, if you can run faster than the others in your group, then you can get away while the zombies are eating them.

There are, of course, other advantages to teaming up. For one thing, there's the comfort of being with other humans. Also, you can keep watch in shifts, and if your team has diverse sets of skills, you can benefit from the expertise of others. Sadly for the old and weak, there is little advantage to their presence over and above being a culinary diversion for zombies, unless they can provide practical skills, wisdom, or leadership.

There is also the possibility that others will be so concerned about their own survival that they will take things from you that they covet. Taking and betraying happen in the best of times, let alone during a zombie apocalypse, so choose your friends wisely. As long as there is mutual advantage in staying together, the group will hold. Generally, the longer you stay together, the more group loyalty will develop as friendships strengthen.

PARTS FOR PROJECTS

This is a project book, so you are going to need parts. Fortunately, a lot of useful material will be littering the streets and roads.

CARS

Car batteries are particularly useful. In fact, cars are full of useful things that can be repurposed (Figure 1-4).

- Horns that can be used as alarms or zombie distractors (see Chapters 4 and 9)
- Alternators for making generators (see Chapter 2)
- 12V light bulbs for illumination and to serve as indicators
- Assorted switches
- Relays for automated switching
- Miles and miles of copper wire

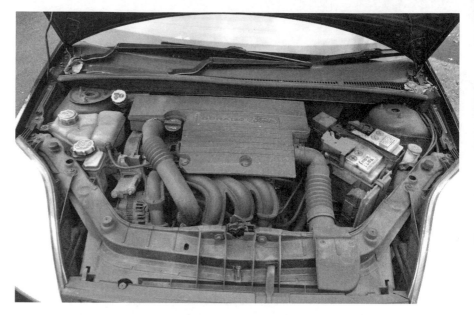

FIGURE 1-4: CARS ARE FULL OF USEFUL STUFF!

Of course, removing the parts from a car out in the open is risky. Have all the tools you need with you and work quickly. If you break into a car, the alarm may sound. It's much better to use a car whose doors are already open.

An alternative to taking things off a car is to just visit an auto parts store or auto mechanic. In fact, if you are trying out some of these projects pre-apocalypse, then visiting a junkyard or auto shop is your best option.

BRICK-AND-MORTAR PARTS STORES

Your hometown probably has a Fry's or some other store from which it is possible to buy (or postapocalypse, take) electronics components. In the UK, Maplin fills a similar niche. While you won't build much from scratch in this book—instead, you'll learn to reuse everyday household items whenever possible—you can get a few really useful things at such stores:

- Walkie-talkies
- Batteries
- Solar panels
- Tools
- Prototyping platforms, such as Arduino and Raspberry Pi controllers (See "Electronic Modules" on page 17.)

Of course, preapocalypse, you could also just order most materials on the Internet. (Then, you'll even be prepared with a stockpile; imagine the bartering possibilities!) See Appendix A for a detailed breakdown of the electronic parts you'll need for this book and where to buy them.

PROJECT CONSTRUCTION

The projects in this book are mostly concerned with the use of electronics in some way. They're all described step-by-step, and no electronics expertise is required. You'll find detailed lists of the supplies you need to make a project within the project itself, and you'll need just a few basic tools, including a soldering iron.

SOLDERING

You use a soldering iron to melt solder, which is used to join wires together or attach components to a circuit board. The basic principal is that you touch the hot tip of the iron to the solder, without burning yourself. During the apocalypse, burn medication will be hard to come by, so take care.

There is, of course, the problem that we might not have a supply of electricity to power the soldering iron. Fortunately, several types of cordless soldering irons are available. There are butane gas–powered irons, as well as irons that are battery powered. You can even repurpose a hot plate or a toaster oven to solder components to circuit boards.

In Appendix B, you will find a beginner's guide to soldering. Trust me: if you can use a knife and fork, you can solder.

MECHANICAL CONSTRUCTION

You'll want to put the contraptions you make in this book into boxes or affix them to walls, so it will be helpful to get hold of a drill as well as screws, nuts, bolts, and metal brackets. General construction tools such as a hacksaw, files, and a vise will come in very handy for fashioning supports and fixings from scrap metal or lumber. The more tools you can lay your hands on, the better. They can always double as weapons.

ELECTRONIC MODULES

Wherever possible, the projects in this book use ready-made modules to simplify the build. Two such modules are the Arduino (Figure 1-5) and Raspberry Pi (Figure 1-6). You will find a guide to the Raspberry Pi in Chapter 5 and an Arduino primer in Appendix C.

FIGURE 1-5: AN ARDUINO MICROCONTROLLER BOARD

FIGURE 1-6: A RASPBERRY PI SINGLE-BOARD COMPUTER

The Arduino is a microcontroller board widely used by makers and artists. It's simple to use and can be programmed to read sensors and control outputs. For example, in Chapter 2, you will use it to make a battery monitor, and in Chapter 9, you'll use it to control an LED flashlight to make an automatic Morse code beacon.

The Raspberry Pi is a much more sophisticated device. It is a low-power computer running the Linux operating system. You can connect a keyboard, mouse, and a TV to it and turn it into a control center for your base. Being low power, it is much more suitable to run on batteries than a laptop computer would be.

If you are new to programming, don't worry: all the program code for the projects that use Raspberry Pi and Arduino is available for download from *http://nostarch.com/zombies/*. You may want to download the code to a pen drive now, just in case.

In the next chapter, I will start with the basics of sorting out the electrical power that you will need for most projects. Having electricity available will make life easier in other ways, too, such as by providing lighting, so let's get started!

2

GENERATING ELECTRICITY

In the aftermath of a zombie apocalypse, the national power grid is likely to continue working for only a day or two at most. The system of power generation and distribution is finely balanced and fantastically complex, and the people who run it are likely to be busy either being eaten by zombies or being zombies (Figure 2-1), so you won't be able to rely on them.

Let's face it, though: you're not going to need to have a whole lot to power, anyway. There won't be any TV to watch, and you won't have the Internet, either. You'll only need a fairly small amount of electricity, and fortunately, you can generate that much yourself, either by using the sun's energy or by converting movement into electricity.

FIGURE 2–1: ZOMBIE WORKERS

POWER AND ENERGY

The words *energy* and *power* are often used interchangeably, but they're actually different. *Power* is the amount of *energy* used per unit time, usually per second. Energy is measured in units called joules (after James Joule, the English scientist and brewer). You could represent power in units of joules per second, but power is more commonly measured in watts (named for James Watt, the Scottish inventor). One watt is actually exactly one joule per second.

Think of a battery as holding a certain number of joules of energy. How fast the battery empties depends on how much power you draw from it. If you attach a very low-power device, the battery will take a long time to go dead, but if you attach something high power, the battery won't last long at all.

Table 2-1 lists some electrical appliances and indicates just how much power they use.

Table 2-1: POWER CONSUMPTION OF EVERYDAY ITEMS

APPLIANCE	POWER (W)	WOULD DRAIN A CAR BATTERY IN:
Portable FM radio	2	300 hours
LED light bulb	5	120 hours
Soldering iron	30	20 hours
Laptop	50	12 hours
Monitor (27 inch)	80	7.5 hours
Hair dryer	1,500	24 minutes
Electric room heater	3,000	12 minutes
Electric shower (the type that both pumps and heats the water)	10,000	3.6 minutes

Cooking and heating require a lot of power. In fact, if you want hot water or hot food, you should look at burning fuel rather than using electricity.

FLAVORS OF ELECTRICITY

Although Table 2-1 lists a portable radio and an electric shower, these things need different types of electricity. Fortunately, that doesn't have to be a problem! With some constraints, it is possible to convert between these types. Note that zombies will be unable to manage this task (Figure 2-2).

Devices that use electricity fall into two categories: those that require high-voltage alternating current (AC) and those that require low-voltage direct current (DC). DC devices are often battery powered.

FIGURE 2-2: FLAVORS OF ELECTRICITY

LOW-VOLTAGE DC

Low-voltage DC is much safer and easier to generate, use, and store than AC. *Low-voltage* generally means 12V (volts) or less. I find it helps to think of water flowing through pipes when trying to understand how electricity flows through wires. This image is particularly useful for understanding the difference between *voltage* and *current*.

Voltage is like the pressure in a water pipe. A high voltage can supply much more power than a lower voltage can, just as a high-pressure pipe could fill a container with water much faster than a lower-pressure pipe could. But thinking of voltage as just pressure creates an incomplete picture; it's more accurate to think of voltage as a height difference.

In the schematic (Figure 2-3), the point where the water enters the pipe is above the point where the water leaves the pipe. The higher the entrance is above the exit, the greater the rate of flow. This rate of flow is called the current, and in electronics, the current is the amount of charge passing a point per second. The unit of measurement for current is the ampere, which is abbreviated to just *A*. It is also common to see current measured in *mA* (milliamps). One mA is 1/1,000 of an A.

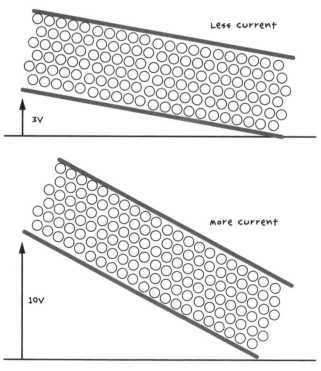

FIGURE 2–3: VOLTAGE AND CURRENT

Interestingly, you can work out the amount of power that something uses by multiplying the power in voltage (V) by the current in amperes (A).

When supplying some low-voltage equipment (let's say an FM radio receiver) with power, it's important to get the voltage correct. Too much voltage will cause too much current to flow through the radio and may kill it. The last thing you need is a zombie radio! Similarly, if there's too little voltage, not enough current will flow to make the thing work properly. The range of acceptable voltage can be quite wide, depending on the device. For example, a radio indicated as requiring 6V to operate may work perfectly well at anything between 4V and 8V.

WARNING When using a low-voltage DC device, make sure you put the batteries in the right way around. Batteries have a positive and a negative connection. If you connect them incorrectly, the current will try to flow the wrong way through the device. If the device does not have internal protection against this (note that most do), the device may be rendered nonfunctional.

HIGH-VOLTAGE AC

High voltage is used to distribute electricity to people's homes because higher voltage makes power transmission more efficient. High-voltage AC is very different from low-voltage DC. For one thing, the voltage is either 120V (in the United States) or 220V (in most of the rest of the world). Also, AC voltage is alternating: unlike a battery, which has one positive connection and one negative, an alternating current switches the polarity of its two leads between positive and negative at a rate of 60 times a second (in the United States) or 50 times per second (in most of the rest of the world). The unit for frequency, which is the number of times that the electricity switches polarity per second, is hertz (Hz).

How the voltage changes over time with an AC power source can be graphed, as in Figure 2-4. Notice that the voltage doesn't suddenly switch direction but rather swings gently one way and then the other, gradually increasing to a peak of over 150V and then down to below −150V. Clearly, this is more than 120V on either side of zero. The maximum and minimum are described as 120V because this amount of AC power provides the equivalent amount of power as 120V DC. This way of measuring AC voltage is called root mean square (RMS). For more information on this topic, take a look at *http://www.electronics-tutorials.ws/accircuits/rms-voltage.html*.

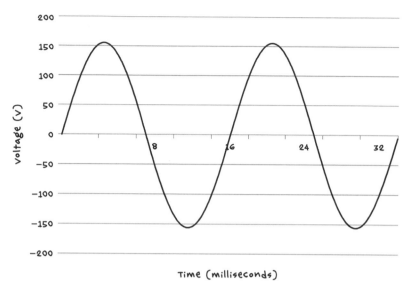

FIGURE 2-4: ALTERNATING CURRENT (AC)

Low-voltage DC devices are often run on AC by using an adapter, like the one your laptop uses or the "wall wart" that you plug your phone into, which converts the AC into DC and drops the voltage at the same time. In our postapocalypse world, unless you have an AC generator, you're likely to be both making and using low-voltage DC directly. Although you can convert DC to AC with a device called an inverter, converting in either direction is inefficient, wasting some energy, and is best avoided.

If you decide to use an inverter, remember that even though you are powering it from a battery, it is generating high and therefore dangerous voltages. Therefore, exercise the same common sense as you do when plugging devices into an AC wall outlet.

BATTERIES

Batteries, used to store electrical energy, come in lots of different types. Some are small and single-use, like AA cells. Others, such as lithium laptop batteries and lead-acid car batteries, can be recharged. Note that batteries only supply DC.

Both single-use and rechargeable batteries are essential to your survival during a zombie apocalypse, so scavenge as many as you can during your supply runs. As you'll see in Chapters 9, 10, and 11 of this book, you can use batteries to power zombie-distracting devices and communications devices. Of course, both types of batteries have different merits. Let's explore those now so you can decide which deserves a spot in your go bag.

SINGLE-USE BATTERIES

AA batteries have a long shelf life, and to operate many small appliances, it makes sense to scavenge a good supply of these. They also run out of power slowly. For example, if your flashlight begins to dim, you'll still get a few valuable minutes of light before the battery completely dies. Note that rechargeable AA batteries usually give out much more quickly than single-use batteries—and with less warning.

RECHARGEABLE BATTERIES

Lithium polymer (LiPo) batteries have transformed mobile devices because they're lightweight and can store a lot of energy. Since a cellphone is so easy to carry around, you might think LiPo batteries are a good rechargeable choice for any portable postapocalyptic device. But be warned: they have a few quirks:

- They are prone to catching fire if overcharged, punctured, or cut.
- They require special charging circuits.
- They don't work well at extremes of temperature.

In short, for storing energy that you generate, it's better to use the lead-acid batteries that you find in cars. For a start, there should be a plentiful supply of these. They also have the advantage of working at low temperatures, and they are much more forgiving of overcharging or continuing to be discharged after they are empty than other types of rechargeable batteries. The only real downside to lead-acid batteries is that they are really heavy, so when you need to scavenge car batteries, don't be tempted to load your pack with much else. Otherwise, you'll quickly find yourself too overburdened to escape a pursuing zombie.

BATTERY CHARGING

Under normal circumstances, the easiest way to charge a battery is to use an AC-powered battery charger. Since you won't have access to AC (unless you've hit the jackpot and found a working generator), you need to consider ways you can generate electricity to charge batteries.

In the project that follows, you'll learn how to generate electricity and charge batteries using solar power, in many ways the easiest solution to postapocalyptic power problems. You'll then discover how a stationary bicycle and a car alternator can be adapted to charge batteries. The principles you learn here also govern using water wheels and wind turbines. In fact, anything that can turn the shaft of a car alternator at a reasonable speed and with reasonable force can be used to generate power. A drive belt is a good way to link whatever is turning to the alternator and provide some gearing so that the alternator moves fast enough.

PROJECT 1: SOLAR RECHARGING

This project will show you how to make a simple setup that charges a 12V car battery using solar power.

SOLAR PANELS

Photovoltaic (PV) solar panels are silent, require minimal maintenance, and will just sit there happily generating electricity. They generate a lot more electricity when the sun is out, but they still make useable amounts of electricity on an overcast day. Obviously, they're useless at night, which is why you'll use them to charge batteries, not power a device directly. They also need to be situated with a clear view of the sky and out of zombie climbing height, as an undead entity partially obscuring a solar panel will drastically reduce its efficiency.

You may find solar panels to scavenge on the roofs of houses or even in arrays on the ground. Your electricity needs are likely to be relatively modest, so one or two panels will be plenty. After all, we're talking about survival here; the hot tub can wait.

As you might expect, the generating capability of solar panels is measured in watts. But make no mistake: a solar panel labeled "100W" may generate just about 100W at noon on a cloudless day on the equator, but most of the time, it will generate a lot less than that.

Solar panels incorporate different types of technology, the most common types being *monocrystalline silicon* and *polycrystalline silicon*. The mono panels are more efficient and produce more power per square foot, but the poly panels still make perfectly good electricity. They just need to be a little larger to make as much. It does not matter what type you take; all you really need to be aware of is the number of watts. If you turn the solar panel over, you should find a label that gives you all the key data about the panel.

CHARGE CONTROLLERS

Domestic solar installations don't charge batteries. Instead, a complex piece of equipment converts the low-voltage DC produced by the solar panels into high-voltage AC. The converted power is first used to meet the demands of the house's AC wall sockets and lighting. Then anything left over goes into the power company's AC lines, and the power company pays for the contribution of excess electricity. Well, that's what happens if you're reading this before the apocalypse. Otherwise, it's likely everyone at the power company has become a zombie, and money has become meaningless.

Instead of giving your excess electricity away to a power company that doesn't care and won't pay for it, store it in batteries for later use. This project works just like the way you'd manage electricity for a motor home or boat that uses PV cells to charge its batteries when the vehicle is not in use.

Rather than build an electronic circuit to control the charging, it's much easier and more reliable to use a ready-made *charge controller*. If you're buying preapocalypse, then pick one up on eBay, at another online retailer, or at a physical store, like Fry's. If you're buying postapocalypse, then they're available free of charge from physical stores.

WHAT YOU WILL NEED

To make this project, you'll need the following items.

ITEM	NOTES	SOURCE
☐ charge controller	7A (or more) 12V	eBay, Fry's (4980091), abandoned RVs and boats
☐ PV solar panel	20W–100W	eBay, scavenge
☐ car battery	12V	Auto parts store, scavenge
☐ 2x heavy-duty alligator clip	7A or more	Auto parts store
☐ Electrical cable	7A	scavenge
☐ Terminal block	10A	Home Depot, Lowe's, Menards
☐ Multimeter	Simple multimeter	Auto parts store, eBay, Fry's

Solar panel specifications have become pretty standardized. Look for a solar panel that generates between 20W and 100W and is labeled as 12V. That means the panel is suitable for charging 12V batteries. Nominally, 12V solar panels will actually produce upwards of 18V.

The power cable needs to be long enough to connect the solar panel to the charge controller. This cable could be an AC outlet extension with the connectors cut off each end. Thin, low-current cable has a higher resistance to the flow of current than higher-current cable, which will waste precious power. For example, a 10A AC outlet extension cable that's 30 feet (10 m) long will waste about 0.5W of power for a 20W solar panel charging at about 12W. For this reason, use a thick cable and keep its length short if you can.

Since you will be making your multimeter a permanent part of this project and you will also be chopping up the test leads, I urge you to use the cheapest possible multimeter. You will probably also find it useful to have another multimeter to use for testing.

In addition to the components listed above, you will need the following general construction tools:

- Drill
- Screws (assorted sizes)
- Screwdriver

You're going to use multimeters a lot in this chapter, too. Take a look at "Using a Multimeter" on page 237 to find out more about how to use this useful little tool.

CONSTRUCTION

The most difficult part of this project is likely to be fixing the solar panel somewhere reliable, where the zombies and wind cannot displace it. A roof is probably a good idea, but it's up to you to figure out the best place for the panel in your compound. Remember, you're going to need to run a cable from the solar panel to the area where you plan to keep the battery and charge controller.

The diagram shows the wiring for the project (Figure 2-5).

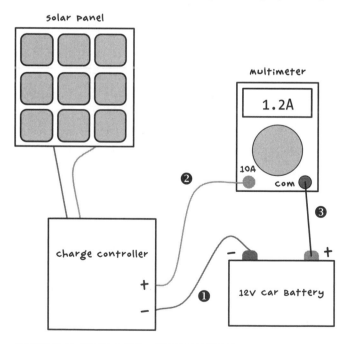

FIGURE 2-5: SOLAR PANEL WIRING DIAGRAM

Charge controllers are all a little different from each other, but most will have six terminals, each in pairs of + and–. One pair will connect to the solar panel, another will connect to the battery, and the third pair (not shown in Figure 2-5) will connect to whatever you want to power with the battery. For now, let's just worry about charging the battery; I'll show you how to use that stored energy later.

The charge controller will monitor the battery voltage and the voltage coming from the solar panel to ensure that the battery doesn't overcharge or deplete so far that it stops accepting charge. More advanced models may have a display to show you what's going on, but I used a very basic model, so I also used a multimeter to show how much current is flowing into the battery. If your charging controller does this for you, then you can probably do without the multimeter. In that case, the charge controller's positive connection goes straight to the positive battery connection, without the meter in between.

STEP 1: FIX THE SOLAR PANEL

It should go without saying that the solar panel should go somewhere sunny and far out of a zombie's reach, but near a window inside your base won't be good enough. Ideally, it needs to be on a south-facing roof. The angle depends on your latitude. For optimal performance, the further from the equator you are, the closer to the vertical the panel should be tilted. If your base has a slanted roof, you can probably just attach the panel to the natural slope of the roof, as roofs tend to have steeper angles further from the equator to allow snow and rain to run off more easily.

You may have to improvise with wooden batons to attach the panels. The photograph shows my solar panel mounted on a roof (Figure 2-6).

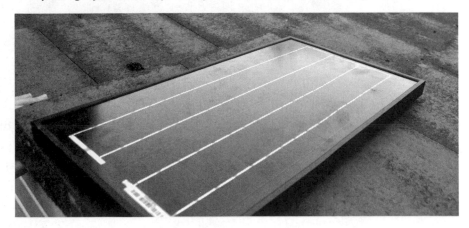

FIGURE 2-6: SOLAR PANEL READY TO MAKE POWER

STEP 2: ATTACH A LEAD TO THE SOLAR PANEL

The solar panel may have screw terminals, or, as mine does, it may have
a short length of wire soldered to its terminals. The lead attached to the
solar panel needs to be long enough to reach inside your base, where you
can ensure it stays dry. Attach a lead that can be fed through a hole on the
wall or in the roof and attached to the screw terminals. Just like zombies,
water is likely to find its way through any gap, so seal up holes after you
have threaded the cable through. Silicone sealant works well for this.

Once the cable is inside, you can use the terminal block to extend it to
the length you need, though a single length of cable without joins will be
most reliable. A terminal block can be used to join the lead from the solar
panel to a longer lead, which will be connected to the charge controller
(Figure 2-7).

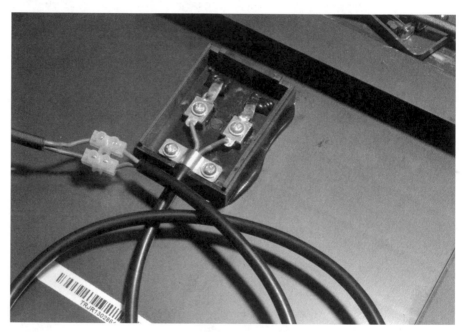

FIGURE 2-7: CONNECTING THE SOLAR PANEL

STEP 3: WIRE UP THE BATTERY AND CHARGE CONTROLLER

Wire up the battery, multimeter, and charge controller as shown in
Figure 2-5. The red probe lead from the multimeter can fit in the screw
terminal of the charge controller, but the black probe needs to connect to
the battery somehow. The best way is to attach one end of the black probe
to one of the heavy-duty alligator clips.

This means we need to make three leads, using the alligator clips and probe leads. For the first of these leads (labeled ❶ in Figure 2-5 and shown in Figure 2-8, I just used half of the multimeter's black lead, with the probe cut off. However, you can use any black wire you like.

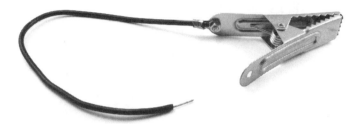

FIGURE 2–8: NEGATIVE BATTERY LEAD

This lead will connect the battery to the negative (–) terminal of the charge controller. To make the lead, strip about half an inch (10 mm) of the insulation from each end of the wire. Connect one of those ends to the alligator clip by wrapping the wire clockwise around the loosened bolt on the clip. Then tighten the bolt so the clip grips the bare wire.

NOTE The wire should be wrapped clockwise around the bolt so that when you turn the bolt, it pulls the wire around with it rather than pushes it away. The connection just works better that way.

Use a pair of pliers to wrap the supporting tabs at the end of the clip around the wire. These will prevent the wire from pulling off the clip if the wire is accidentally pulled on.

The second of the three leads (labeled ❷ in Figure 2-5 and shown in Figure 2-9) will go from the positive high-current terminal of the multimeter to the positive battery output of the charge controller. This lead is just the positive meter lead with the probe cut off and the insulation stripped off the last half inch (10 mm).

FIGURE 2–9: POSITIVE CHARGING LEAD

You'll connect the final of the three leads (labeled ❸ in Figure 2-5 and shown in Figure 2-10) from the negative (COM) connection of the multimeter to the alligator clip that will be connected to the positive terminal of the battery.

FIGURE 2–10: POSITIVE BATTERY LEAD

Strip about half an inch (10 mm) of insulation from the remainder of the black probe lead of the multimeter and attach it to the alligator clip in the same way you did for the lead ❷. This new lead is going to be connected to the positive terminal of the battery.

This lead is black, however, and since the convention is that black means negative, the color could be confusing. To make the purpose of this lead more intuitive, wrap some red electrical tape around it to make red stripes and add some red tape to the "finger end" of the alligator clip.

Now use the three leads to connect everything together, ready for use (Figure 2-11). Note that most multimeters have a special positive socket just for high currents. This may be labeled 10A or 5A. Plug the red lead into that socket. Be sure to set meter to the correct range, which is DC current at the meter's maximum available current reading.

STEP 4: TESTING

To test the solar panel, use the highest DC amps setting of the multimeter to monitor how much current flows into the battery from the panel via the charge controller. If the battery needs charging, the charge controller should attempt to charge the battery as much as possible until it is full, and the meter should show a positive reading. After the battery is full, most charge controllers will switch to a trickle-charge mode that just keeps the battery topped up.

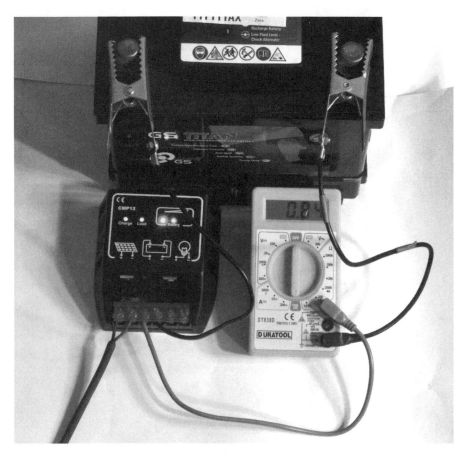

FIGURE 2-11: THE CHARGE CONTROLLER, MULTIMETER, AND BATTERY ARE CONNECTED.

In Figure 2-11, 0.84A of current is flowing into the battery. If your current reading is negative, then current is flowing out of the battery. This means something is wrong, so check over your wiring. You should also see the current drop considerably if you cover part of the solar panel or if the sun goes away.

If the battery doesn't need charging, then the meter should read zero, which doesn't tell you much. You'll have to wait until Chapter 3, when we attach some lighting to the battery, to see the charging process in action.

USING THE SOLAR CHARGER

To ensure a continuous supply of electricity, it's a good idea to duplicate this entire design so that if one solar panel or set of wiring should fail, you have a spare. Since swapping batteries just means unclipping the alligator clips, you can even keep a stack of batteries in rotation and set a few fully charged batteries aside for emergencies. Stockpile batteries for your own base, or start a new career as a postapocalyptic battery shop owner. Money might be useless, but I'm sure you could barter electricity for food, supplies, or assistance with your next scavenging trip.

On a sunny day, if lots of current is flowing from the solar panels, you may find that the alligator clips get hot. Wrapping some tape around them will reduce the chance that you'll burn your fingers when you swap in a new battery.

PROJECT 2: BICYCLE GENERATOR

In this project, you'll generate power with an adapted bicycle, which doubles as a great way to stay in shape so you can outrun the undead. The design uses a car alternator to charge the car battery. The alternator is just doing what it would naturally in a car, but without an engine. The alternator includes all that is needed to charge the car battery, so in this project, there is no need for the charge controller that you used in the solar project.

NOTE You could also adapt this project to use other forms of rotary movement. For example, you could connect it to a wind turbine, a water wheel, or zombies on a treadmill (Figure 2-12).

FIGURE 2-12: ZOMBIE POWER

WHAT YOU WILL NEED

To make this project, you will need the following items.

ITEM	NOTES	SOURCE
☐ Bike	Large wheels	Scavenge
☐ Car alternator	Almost any will work	eBay, scavenge
☐ Car battery	12V	Auto parts store, scavenge
☐ Drive belt	v belt, size A100	Auto parts store, eBay, hardware store, scavenge
☐ 2x heavy-duty alligator clip	7A or more	Auto parts store
☐ Butt terminals for alternator + and −	To suit your alternator terminals	Auto parts store
☐ Spade terminal for F terminal of alternator	To suit your alternator terminals	Auto parts store
☐ Electrical cable	7A	Scavenge
☐ Multimeter	Simple multimeter	Auto parts store, eBay, Fry's
☐ Lamp and holder	12V 5W lamp	Auto parts store
☐ Fuse	10A fuse and holder	Auto parts store
☐ G-clamp		Hardware store
☐ 2 x 4 lumber	5 feet (1.5 m)	Hardware store

This is another project that uses a multimeter as a fixture, and I would again suggest you use the cheapest possible multimeter. You will probably also find it useful to have a spare multimeter to use for testing.

CONSTRUCTION

The trick to building this project is to keep the cycle's back wheel away from the ground. There are two ways to do this. One is to make (or scavenge) a stand designed to allow a regular bike to be used as an exercise bike. This needs to be strong enough to support you when you sit on the bike, so a maintenance stand probably won't be strong enough.

The other approach is to turn the bike upside down. Then you can use the pedals with either your hands or your feet while you sit in a chair near where the handlebars used to be. I used the upside-down bike approach.

ALTERNATORS

IF YOU KNOW A LITTLE ABOUT ELECTRONICS, YOU MAY BE AWARE THAT MOST DC MOTORS CAN BE USED TO GENERATE CURRENT. MOVING A COIL OF WIRE IN A MAGNETIC FIELD (USUALLY SUPPLIED BY A REGULAR MAGNET) WILL CAUSE A CURRENT TO BE GENERATED IN THAT WIRE.

ALTERNATORS, LIKE ALL GENERATORS, OPERATE ON THIS BASIC PRIN-CIPLE, BUT INSTEAD OF A NORMAL MAGNET, THE MAGNETIC FIELD THAT THE GENERATING COIL TURNS IN IS CREATED BY AN ELECTROMAGNET. THE ELECTRO-MAGNET IS POWERED BY THE ALTERNATOR ITSELF ONCE IT GETS GOING. TO GET GOING, THE ALTERNATOR MUST BE CONNECTED TO THE BATTERY; OTHER-WISE, THERE WILL NO MAGNETIC FIELD FOR THE GENERATION TO START. IT'S A CHICKEN-AND-EGG SITUATION: THE ALTERNATOR NEEDS TO BE GENERATING CURRENT IN ORDER TO GENERATE CURRENT.

FIGURE 2-13 SHOWS A SIMPLIFIED SCHEMATIC OF A CAR ALTERNATOR.

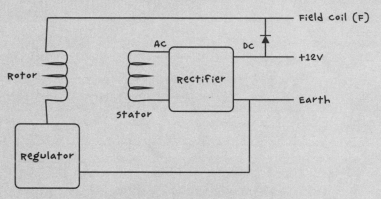

FIGURE 2-13: THE SCHEMATIC DIAGRAM FOR AN AUTOMOTIVE ALTERNATOR

IN ACTUAL FACT, THE ALTERNATOR WILL NORMALLY HAVE THREE STATOR COILS PRODUCING THREE-PHASE AC TO BE TURNED INTO DC (UNLIKE ORDI-NARY DOMESTIC TWO-PHASE AC). IF YOU WANT TO FIND CHAPTER AND VERSE ON ALTERNATORS, THEN TAKE A LOOK AT THE DESCRIPTION AT http://www .allaboutcircuits.com/vol_6/chpt_4/8.html.

STEP 1: MODIFY THE BICYCLE

First, strip off every piece of the bike that you don't need. You can take away the front wheel, both mudguards, and the brakes. The gears can stay.

Remove the rear wheel and remove its tire and inner tube. Then place the drive belt over the wheel and fix the wheel back onto the bike.

STEP 2: FIX THE ALTERNATOR AND BIKE TO THE 2×4

Alternators don't have standard positions for their fixing lugs, so you may have to improvise a little here. The drive wheel of the alternator needs to line up with the cycle's wheel, but the alignment doesn't need to be exact, especially if you use a long belt like the one in my final arrangement (Figure 2-14).

FIGURE 2-14: THE MECHANICAL ARRANGEMENT OF BIKE AND ALTERNATOR

Use a G-clamp to fix the bike to the 2×4 using the saddle. Adjust the saddle first so that it is flat. Alternatively, you could also remove the saddle completely and make a hole of the same diameter as the saddle stem partway through the 2×4.

Where you place the alternator on the 2×4 depends on the geometry of the alternator, the bike, and the drive belt, so I can't give you exact measurements. Fix the alternator in place once the bike is attached to the 2×4 and the drive belt is around the cycle wheel. The alternator I used had a convenient hole that allowed it to be fixed to the side of the 2×4 (Figure 2-15). The drive belt doesn't need to be under a lot of tension, but you can create a simple tensioner with a spring or elastic strap; you can even cut the latter from the discarded bike inner tube.

FIGURE 2-15: ATTACHING THE ALTERNATOR TO THE 2X4

Turn the cycle very gently to make sure everything is working mechanically and then move on to the next step.

WARNING Don't be tempted to try whizzing the alternator around at high speed without attaching the rest of the circuit, because generating high voltage in the coils with no load can damage the built-in electronics of the alternator.

STEP 3: IDENTIFY THE ALTERNATOR TERMINALS

Now that the mechanical part of the generator is built, we can start looking at the electrical side. First identify the connections on the alternator. Although they have slight differences, automotive alternators are remarkably standard, especially those from older cars. Plus, alternators are often easy to remove from older vehicles.

I used a reconditioned alternator that I bought on eBay for just a few dollars (Figure 2-16). Postapocalypse, however, there should be no shortage of abandoned old cars.

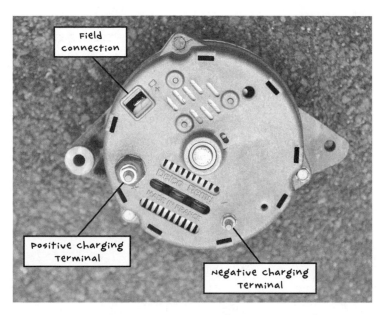

FIGURE 2–16: DELCO LRA443 ALTERNATOR (FROM THE 1980S)

You are looking to identify three connections from the alternator:

Negative charging terminal (–)   This will normally be connected electrically to the metal case of the alternator, but there should also be a bolt specifically for attaching a spade terminal. It may be marked –, GROUND, or GND.

Positive charging terminal (+)   Although it may not look like it, this will be electrically isolated from the metal body of the alternator. This terminal is usually marked with a +, but it may be marked BATT or BATT +. It's quite common for alternators to have two + terminals that are connected together inside the alternator. If this is the case, you can use either terminal.

Field connection (D+)   On my alternator, this is labeled D+, although it is just as common for it to be labeled F.

STEP 4: WIRING

Now that you know which terminal of the alternator is which, you need to make some leads to connect everything together. The wiring diagram shows how to do this (Figure 2-17).

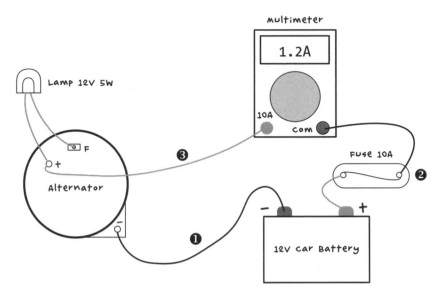

FIGURE 2-17: WIRING DIAGRAM FOR THE CYCLE CHARGER

The light bulb serves two purposes. It limits the current to the field coil so you don't have to pedal too hard to kick-start the alternator into generating. The bulb also serves as a useful indicator: when the alternator starts generating, the light will go out.

There are a few leads to make. Let's start with the lead from the negative terminal of the alternator to the battery negative terminal (❶ in Figure 2-17). This needs a large alligator clip on the battery end and a butt connector on the other, as shown (Figure 2-18).

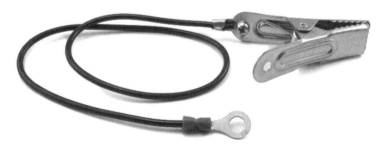

FIGURE 2-18: THE NEGATIVE BATTERY LEAD

You could use any black wire for this lead, but I used two-thirds of a black multimeter lead, from the probe end of the multimeter, and cut off the probe itself. Strip about half an inch (10 mm) of the insulation off each end of the wire. The butt terminal can be crimped (squeezed with pliers) onto the lead. Attach the alligator lead by wrapping the stripped wire clockwise around the bolt before tightening up the bolt.

This charging circuit doesn't include a charge controller to protect it, so you'll need a fuse. *Fuses* are short lengths of metal designed to melt and so break a connection when too much current flows through them. A car battery can store quite a lot of energy—enough to start a fire—so it's worth using a fuse. If something should accidentally short out, the fuse will blow, breaking the connection before too much damage is done.

The most convenient type of fuse holder has trailing wires at each end. You can use these trailing wires to make the lead between the positive terminal of the battery and the multimeter (❷ in Figure 2-17). Attach an alligator clip lead to one fuse wire and attach the remaining third of the black multimeter lead to the other fuse wire. Wrap the connection between the fuse and multimeter leads with electrical tape, and you should have the completed lead (Figure 2-19).

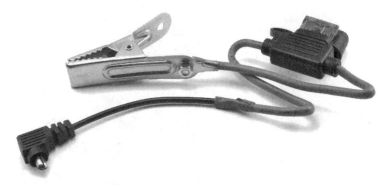

FIGURE 2-19: THE FUSE LEAD

The final of the three leads (Figure 2-20 and ❸ in Figure 2-17) combines both the lamp and the positive charging connector from the alternator to the multimeter.

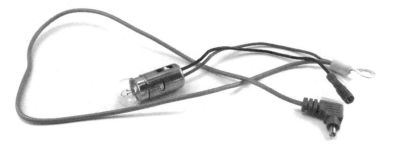

FIGURE 2-20: THE POSITIVE CHARGING AND LIGHT BULB LEAD

As with the fuse holder, I used a light bulb holder with trailing leads. Crimp a ring terminal of the correct size for the F connection of the alternator. It's best to use an insulated spade terminal to minimize the chances of a short circuit.

Chop off the probe from the red multimeter lead, strip half an inch (10 mm) of insulation off the lead, and twist the bare wires together with the other lead of the light bulb holder. Crimp the combined wires together into the butt terminal for the positive charging connection of the alternator.

STEP 5: FINAL ASSEMBLY

With all the leads prepared, it's time to connect everything according to the diagram in Figure 2-17.

The photograph shows the alternator, battery, and multimeter all wired up (Figure 2-21). Before you connect the battery and alligator clip, make sure the multimeter is set to its maximum current range and that the correct sockets for maximum DC current are being used.

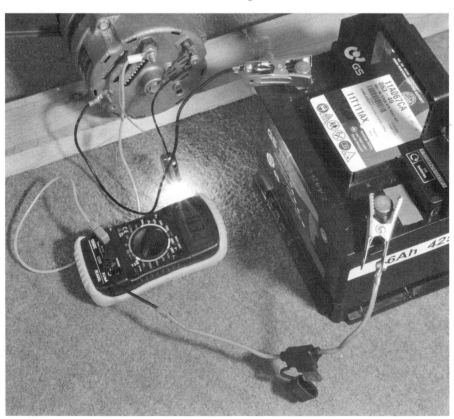

FIGURE 2-21: THE COMPLETED WIRING

USING THE PEDAL GENERATOR

Before you start pedaling, the bulb should be on, and the multimeter should indicate a current of about –0.3A. The value is negative because this current is being used by the lamp.

Crank the pedals quite fast, and you should see the lamp start to dim and then extinguish. At this point, you'll probably feel a lot more resistance from the pedals. This is good news: you're generating electricity! The current should now show a positive value. With furious pedaling, this might increase to 2A or 3A.

If your battery is fully charged, then the bulb should go out while the current remains at zero. This is because the alternator includes a voltage regulator circuit that stops charging the battery when it's up to its maximum voltage. If you need to discharge the battery a little to test that it charges, then you might want to move on to the next project—where we start using some of the energy that we've stored in the battery—and return to test this project later.

You should only connect the charging circuit when you are ready to start using it, as the light bulb will eventually drain the battery completely.

Once you've successfully built one pedal-powered generator, you can repeat these instructions to build a charging station for each person in your base. By generating electricity as a team, you'll stockpile plenty of batteries, and everyone will be able to contribute to your group's continued survival!

We've thoroughly explored a few ways of generating electricity and, in particular, charging up car batteries. In the next chapter, you will learn how to start using this electricity and monitor the state of your batteries so that you don't suddenly get plunged into darkness.

3

USING ELECTRICITY

 Now that you have a neat row of car batteries all charged up and ready to use, it's time to use them to improve your standard of living (see Figure 3-1). First, you'll learn how to connect those batteries to something useful, and then in this chapter's first project, you'll build a simple lighting circuit.

The second project in this chapter will show you how to use an Arduino microcontroller board and a few extra components to make a simple battery monitor. You wouldn't want to lose your brains just because your defenses' batteries died!

FIGURE 3-1: CAR BATTERIES HAVE A VARIETY OF USES IN THE POSTAPOCALYPTIC WORLD.

POWERING DEVICES FROM A CAR BATTERY

Let's look at how you can use all that energy to make your life more comfortable while you keep those pesky zombies at bay. Of course, you first have to get the electricity from the battery to your device. There are two common connectors you'll want to have on hand.

CIGARETTE LIGHTER SOCKETS

As DC voltages go, 12V is pretty useful. It's the same voltage you find in the cigarette lighter socket of a car, and there are lots of 12V appliances that you can just connect straight to the battery. This includes various types of lighting, fans, drink warmers, air compressors, DVD players, mini fridges, and more.

In fact, there are so many 12V appliances with a cigarette lighter plug on the end that it's worth making an adapter lead that will allow you to plug them right in, without having to modify them.

You can buy a cigarette lighter socket adapter, like the one on the left in Figure 3-2, at your local auto parts store. Having acquired it, you can strip the leads and attach alligator clips so you can hook the adapter up to the battery. For info about how to join the wires/attach the alligator clip, see Project 1, "Step 3: Wire Up the Battery and Charge Controller" on page 30.

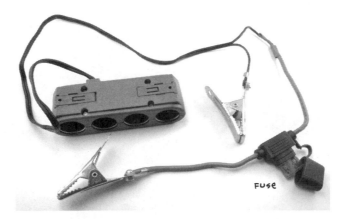

FUSE

FIGURE 3-2: MAKING A CIGARETTE LIGHTER SOCKET ADAPTER

Never assume that a car battery is harmless just because it's only 12V! While you can't get an electric shock from 12V, you can most certainly receive nasty burns from it. If a wrench or screwdriver accidentally shorts across the terminals of a car battery, hundreds of amps will flow through it, turning the tool into flying molten metal that can easily burn or even blind you. Just remember: car batteries store a lot of energy, which can easily be released by such an accident.

Now that you're starting to connect things to your battery, you need to make sure the battery is protected from accidental damage. Some adapters may already incorporate a fuse, but if yours doesn't have a fuse, you should include a fuse holder in the circuit. A 10A fuse, like the one used in "Project 2: Bicycle Generator" on page 34, will work just fine. Always remember to keep spare fuses around. It's not so easy to nip out to the shops if the neighborhood is overrun with zombies. Notice that I've included a fuse in the design in Figure 3-2; the fuse will prevent any problems of a fiery nature, should an accidental short circuit occur.

An alternative (or supplementary) way to protect your battery is to position the solar charge controller from "Project 1: Solar Recharging" on page 26 between the battery and the *load*, or anything you want to power from that battery (Figure 3-3).

When set up this way, the charge controller monitors the battery voltage and will automatically disconnect the load when the battery voltage drops below a certain threshold. This is advantageous because if the battery is discharged too much beyond this point, then it can be so damaged that it will no longer accept charge, and then good luck recharging it.

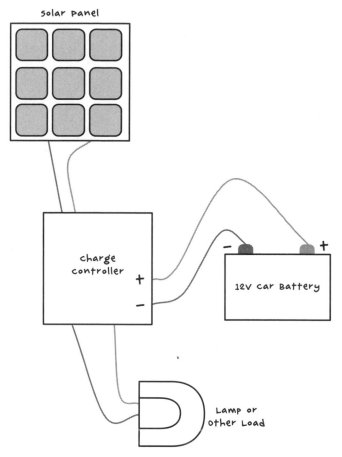

solar panel

charge
controller

+

−

12V car Battery

− +

Lamp or
other Load

FIGURE 3-3: USE A CHARGE CONTROLLER TO PROTECT YOUR BATTERY.
THE SOLAR PANEL IS OPTIONAL.

In "Project 4: Battery Monitor" on page 53, you will learn how to build a battery monitor that will alert you when your battery runs low.

USB POWER

Largely due to the influence of the USB charger lead, 5V has become the most common operating voltage for small DC devices. Jumping from 12V to 5V is much easier than from 120V AC to 5V DC. At the auto parts store, you will find 12V cigarette lighter to 5V USB power adapters.

The adapter in Figure 3-4 has the benefit of combining both 12V sockets and 5V USB sockets. Another type of adapter just has a 12V cigarette lighter–shaped plug with one or two USB sockets built into the end of the plug. You could either plug the type of adapter shown in Figure 3-4 into the adapter of Figure 3-2. Alternatively, you could chop off the plug from the adapter of Figure 3-4 and add alligator clips to it, just as you did to the adapter in Figure 3-2.

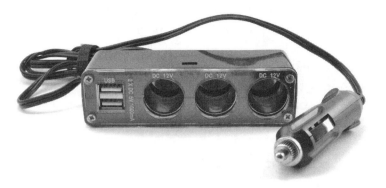

FIGURE 3—4: 12V TO USB ADAPTER

You could, of course, use this adapter to charge your cell phone. But the cellular network will probably be one of the first services to collapse during the zombie apocalypse, first suffering overload from callers flooding the network, trying to get in touch with loved ones, and then succumbing to system failures due to power outages and lack of maintenance.

AC INVERTERS

It's possible to convert the 12V DC of a battery into 120V (or 220V) AC using an *inverter*. This device has terminals that connect to a 12V battery and an AC outlet, which you can plug regular AC appliances into.

You cannot, however, plug in very high-powered AC devices. A wattage rating printed on the inverter will specify the maximum power load it can handle. Small models intended for powering laptops may only be 50W, but 200W or 400W inverters are neither hard to find nor particularly expensive.

Where possible, using DC devices is much better, as inverters aren't very efficient. They generate high-voltage AC and waste quite a lot of energy as heat; just look at the large heat sinks on the sides of most inverters. They also often use significant amounts of current even when nothing is plugged into them, so you have to remember to turn them off when not in use.

In the next section, you'll learn how to make a low-voltage lighting setup that can provide lighting intensity similar to that of AC lighting but by using 12V DC lamps powered directly by a car battery.

PROJECT 3: LED LIGHTING

LEDs offer the most light per watt of any type of illumination and are a natural choice for postapocalyptic lighting. This project uses three 12V, MR16 LED light bulbs. These bulbs are available in powers from 2W to 10W or more, and any of these wattages are suitable for this project.

You can string together more than three of these bulbs if you like. Just use a longer lead and more bulbs. You may, for example, have a long corridor that you wish to defend, and good illumination is essential for effective zombie fighting.

In fact, almost any 12V light source could be used, including high-power halogen light bulbs of 50W or more. However, the higher the wattage, the faster the battery will drain.

WHAT YOU WILL NEED

To make this project, you'll need the following items.

ITEM	NOTES	SOURCE
☐ car Battery	12V	Auto parts store, scavenge
☐ 2x heavy-duty alligator clip	7A or more	Auto parts store
☐ LED lamps MR16	12V 2W–10W	Hardware store
☐ MR16 lamp sockets	Sockets with trailing leads	Hardware store
☐ Switch	Inline switch (5A)	Hardware store
☐ Electrical cable	7A	Scavenge
☐ Fuse	10A fuse and holder	Auto parts store

CONSTRUCTION

The LED lights for this project are wired in parallel (Figure 3-5). In this arrangement, each of the lights gets the full 12V from the battery, and if one of the lights fails for any reason, the other lights will keep working.

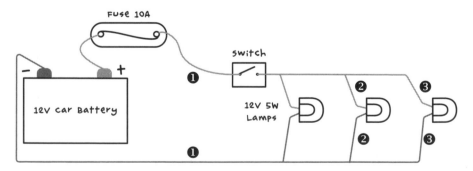

FIGURE 3–5: 12V LIGHTING SYSTEM

STEP 1: PREPARE THE ELECTRICAL CABLE

If you're using a fairly small number of low-power LED light bulbs (up to five, at up to 5W each), then double-core bell wire will be just fine. Cable designed for speakers is also a good choice. In my design (Figure 3-5), I used three bulbs, so I cut three lengths of cable, stripping half an inch (15 mm) of the insulation off the ends.

The first of these leads (labeled ❶ in Figure 3-5) will go from the battery's positive terminal to the switch and then to the first lamp. The second length will continue on to the second lamp (❷), and the final lead (❸) will go to the last lamp.

Where the wires join, there needs to be a three-way twisting of each wire of the cable with both the next length of cable and the lamp holder (Figure 3-6).

FIGURE 3-6: CONNECTING THE LAMP HOLDER

For a more permanent connection, solder the twisted connection. Whether you solder them or not, wrap the connections in electrical tape. For a guide on how to make this kind of twisted-wire connection, see "Joining Wires by Twisting" on page 229.

STEP 2: WIRE UP THE FUSE AND SWITCH

Complete the wiring by attaching the fuse and alligator clip to the start of the wiring. Note that these MR16 light bulbs include a circuit to automatically switch the polarity of the LED. This means you can connect them either way around. If you use a different type of 12V LED, check whether it has separate positive and negative connections. If so, make sure that you connect the positive side to the switch lead and the negative side to the lead you'll attach to the negative terminal on the battery.

The alligator clip attaches to the fuse lead, which is attached to the switch (Figure 3-7). Once again, these leads can be twisted together, and for a more reliable finish, you can solder the twisted joint. The in-line switch uses screw terminals to attach the wires. One side is just a metal connector that passes straight through, and the other side has spring contacts, which connect with each other when the switch is in the on position.

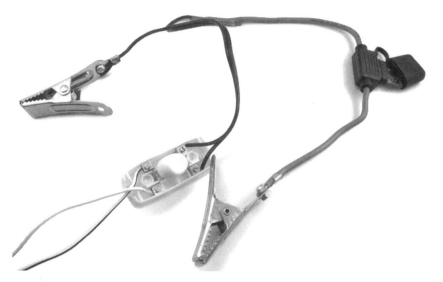

FIGURE 3-7: CONNECTING THE FUSE AND ALLIGATOR CLIP FOR A LIGHTING SYSTEM

STEP 3: INSTALL THE LAMPS

Now that everything is wired up, attach the alligator clips to the battery and make sure that the bulbs light when the switch is flicked. Once you know the lights turn on, just affix them to the ceiling, the wall, or wherever you want them.

USING THE LIGHTING

Murphy's law dictates that batteries will run out of juice and shut off the lights just as the zombies attack. To anticipate and avoid this situation, it's good to know roughly how many hours of light you're going to get from your battery before you need to do some more pedaling or otherwise put some juice into it.

That number of hours depends on the size and quality of your battery. Looking back at Table 2-1 on page 21, you can see that a 5W LED is expected to last 120 hours. Therefore, a string of six 5W LEDs should be good for about 20 hours. If you go all out and put up a string of three 60W 12V halogen lamps, you'll only get about 4 hours of light before needing another charge.

Whatever your lighting setup, it would be great to have advance notice that the battery is getting low—and this is the goal of the next project.

PROJECT 4: BATTERY MONITOR

I recommend keeping a good stock of car batteries charged up and ready to go at all times. That way, if zombies damage your solar panels, or you fall ill and can't pedal your generator, you won't be plunged into darkness and left powerless (in both senses of the word). It is therefore of paramount importance that you have an early warning system that will monitor the battery, telling you when it starts to get low so you can swap in a new one.

This project uses an Arduino, a useful little board that's great for putting together electronic projects that require a bit of logic. In this case, the logic is simply to measure the battery voltage, display it, and sound a buzzer when it falls below a certain critical level.

The Arduino will be powered from the same car battery that it's monitoring. The Arduino uses less than 1W to operate, so it's okay to leave the board connected to the battery continuously.

In the battery monitor setup (Figure 3-8), alligator clips connect the battery monitor to the battery. If the battery has large alligator clips attached to it, then these smaller clips can be attached to the handles of the big clips.

FIGURE 3-8: BATTERY MONITOR

The left lead of the left resistor (Figure 3-8) is connected to the positive battery terminal, and the right lead of the right resistor is connected to the negative battery terminal.

WHAT YOU WILL NEED

To make this project, you'll need the following items.

ITEM	NOTES	SOURCE
☐ Arduino	Arduino Uno R3	Adafruit, Fry's (7224833), Sparkfun
☐ Arduino screwshield	screwshield	Adafruit (196)
☐ LCD shield	LCD 16x2 display shield	eBay, Sparkfun (DEV-11851)
☐ Buzzer	small piezo buzzer	Adafruit (1740), eBay
☐ 270Ω resistor		Mouser (293-270-RC)
☐ 470Ω resistor		Mouser (293-470-RC)
☐ small alligator clip leads		Auto parts store

One great thing about using an Arduino is that there are many different ready-made modules, called *shields*, that fit on top of the Arduino and add extra features to it without any complex electronic construction. This project uses two such shields that are stacked on top of each other.

The first shield that fits on top of the Arduino is a *screwshield*, sometimes called a *wing shield*. This shield allows you to attach wires to the Arduino using screw terminals and a screw driver. The second and topmost shield that you'll attach to the Arduino is an LCD display shield. This shield will tell you the battery level as a measurement of voltage and as a bar graph display of the state of charge (SOC) of the battery. The project also has an option to mute the buzzer to avoid attracting zombies, if you suspect they are shuffling about nearby.

The only other electronic components in this project are a pair of resistors and a buzzer. The resistors are needed because although the Arduino has inputs to measure voltage, it can only measure voltages up to 5V. Any more than that would damage the Arduino. You'll use the pair of resistors in an arrangement called a *voltage divider*. The resistors I've chosen for my divider reduce the voltage to the Arduino by a factor of 2.74 so that the 12V or 13V that we might find at the battery will be reduced to 4.7V or less.

VOLTAGE DIVIDERS

USING TWO RESISTORS AS A VOLTAGE DIVIDER (FIGURE 3-9) IS A GREAT WAY TO REDUCE THE VOLTAGE YOU ARE TRYING TO MEASURE TO A LEVEL WHERE IT CAN BE DIRECTLY MEASURED BY, SAY, AN ARDUINO.

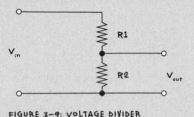

FIGURE 3-9: VOLTAGE DIVIDER

THE FORMULA TO CALCULATE V_{OUT} IF YOU KNOW V_{IN}, R1, AND R2 IS

$$V_{OUT} = V_{IN} \times \frac{R2}{R1 + R2}$$

FOR EXAMPLE, IF R1 IS 470 Ω, R2 IS 270 Ω, AND THE MAXIMUM VOLTAGE OF V_{IN} IS 13V, THEN

$$V_{OUT} = 13V \times \frac{470\Omega}{470\Omega + 270\Omega} = \frac{3510}{740} = 4.74V$$

IN OTHER WORDS, EVEN IF YOUR BATTERY IS FULLY CHARGED AND MANAGING TO PROVIDE 13V, ONLY A MAXIMUM OF 4.74V (BELOW THE CRITICAL 5V LEVEL) WILL FIND ITS WAY TO THE ARDUINO. IF THE INPUT VOLTAGE IS LOWER THAN THIS, THEN V_{OUT} WILL SCALE PROPORTIONALLY. FOR EXAMPLE, IF THE BATTERY VOLTAGE IS 6.5V (WHICH WOULD INDICATE A BIT OF A PROBLEM, BY THE WAY), V_{OUT} WOULD BE 2.37V.

CONSTRUCTION

Remarkably, no soldering at all is needed to make this project. The only tool you need is a screwdriver.

STEP 1: PROGRAM THE ARDUINO

Arduino programs, which are called *sketches*, can change whether a connection, or *pin*, on the Arduino is an input or an output. The Arduino remembers whether each pin was set to input or output, even after you disconnect it from the rest of the circuit. Thus, if one of your Arduino pins was an output the last time you used it, connecting the Arduino to new hardware that expects the pin to be an input could damage the Arduino or the circuit you're connecting it to. By uploading the program to the Arduino before doing anything else, you'll ensure that each pin functions the way your circuit expects it to.

You'll find detailed instructions on getting started with the Arduino, connecting it to your computer, and uploading a sketch to it in Appendix B. In this case, the sketch is called *Project_04_Battery_monitor* and can be found with all the other program files used in this book at *http://nostarch.com/zombies/*.

STEP 2: BUILD THE ARDUINO SANDWICH

When used with the two shields, the Arduino Uno is on the bottom, the screwshield is plugged into that, and, finally, the LCD display shield goes on top of the screwshield (Figure 3-10). The LCD shield has to be at the top of the stack or you won't be able to see what it says!

FIGURE 3-10: AN ARDUINO "SANDWICH"

When pushing the pins of a shield into an Arduino or the screwshield, be careful to check that all the pins meet the holes correctly so you don't damage them. It's quite easy for one of the pins to splay out as you are pushing the pins in.

STEP 3: ATTACH THE RESISTORS AND BUZZER

You'll attach the resistors and buzzer to the screw terminals of the screwshield (Figure 3-11).

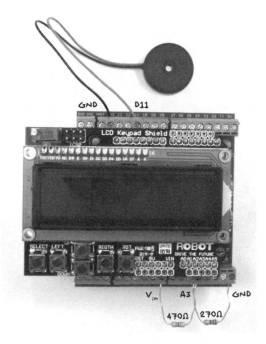

FIGURE 3-11: CONNECTING COMPONENTS TO THE SCREWSHIELD

The two resistors can be identified either by measuring their resistance using a multimeter (see "Using a Multimeter" on page 237) or by reading the colored stripes on the resistor body. The 470 Ω resistor will have stripes of yellow, purple, and brown; the 270 Ω resistor will have red, purple, and brown stripes. In "Resistor Color Codes" on page 225, you will find a resistor color code table and instructions on how to identify resistors by their stripes.

Some buzzers will have a positive red lead and a negative black lead. If this is the case, connect the black lead to GND (ground) and the red lead to D11 on the Arduino. Other buzzers will have identical leads; if this is the case, it doesn't matter which way around they are connected.

SOFTWARE

The sketch for this program is mostly concerned with making sure that the right text is displayed on the LCD at the right time. I'll walk you through it in full, though you don't have to understand or follow how this sketch works to finish the project. You can just upload it exactly as it is into the Arduino board, following the steps explained in "Installing the Antizombie Sketches" on page 248.

If you want to learn more about Arduino programming, see Appendix C or take a look at my book *Programming Arduino: Getting Started with Sketches* (McGraw-Hill, 2012).

The sketch starts by importing the *LiquidCrystal* library, which is responsible for controlling the LCD shield. Because this library is included as a standard part of the Arduino software, there is nothing to download and install for this library.

```
#include <LiquidCrystal.h>
```

After the library command, three constants are defined for key battery voltages.

```
const float maxV = 12.6;
const float minV = 11.7;
const float warnV = 11.7;
```

These voltages are, in order, the fully charged battery voltage, the minimum voltage that you want the battery to be allowed to discharge to, and the voltage at which the buzzer should sound. These last two are both set to 11.7V. These values are pretty standard for a lead-acid car battery, but if you use a different type of battery, you can tweak them. Because they hold decimal values, the variables are of a type called *float*. You can find out more about Arduino data types in Appendix C.

The next few lines define constants for the Arduino pins that are used.

```
const int buzzerPin = 11;
const int voltagePin = A3;
const int backlightPin = 10;
const int switchPin = A0;
```

The Arduino's various pins are normally identified simply by a number, so these constants give them meaningful names. You don't need to change these pin designations unless you decide to wire up your battery monitor differently.

The final section defines constants for the values of the resistors used in the potential divider.

```
const float R1 = 470.0;
const float R2 = 270.0;
const float k = (R1 + R2) / R2;
```

The constant k is the resulting factor the input voltage will be reduced by in order to fit into the 5V measurement range of the Arduino. The next line of code initializes the LCD display, specifying which pins are used.

```
//                RS,E,D4,D5,D6,D7
LiquidCrystal lcd(8, 9, 4, 5, 6, 7);
boolean mute = false;
```

The comment line starting with // just identifies which of the Arduino pin numbers on the line beneath it correspond to which pins on the LCD module. The line after that defines a *Boolean* (a value that can be true or false) variable, mute, which is used to mute the buzzer.

The setup function that comes next is run just once, when the Arduino starts. In this case, it begins by setting the backlight pin (D10) to be an input.

```
void setup()
{
  // Because of a defect in common cheap LCD displays,
  // backlight controlled by transistor D10 high can
  // burn out Arduino pin
  pinMode(backlightPin, INPUT);
  lcd.begin(16, 2);
  lcd.setCursor(0, 0);
  lcd.print("Battery ");
}
```

The backlight pin is used only on some LCD shields, but a significant number of LCD shields have a design flaw that can destroy the Arduino they are connected to if this pin is set to an output and also set high. To be on the safe side, D10 is set to an input. The rest of the function initializes the LCD display and writes out the word Battery, which will be a permanent fixture of the message displayed.

The loop function that follows the setup function is run repeatedly. That is, as soon as all the commands in the function have been executed, the function will start again from the top.

```
void loop()
{
  displayVoltage();
  displayBar();
  if (readVoltage() < warnV && ! mute)
  {
    tone(buzzerPin, 1000);
  }
```

```
  if (analogRead(switchPin) < 1000) // any key pressed
  {
    mute = ! mute;
    if (mute) noTone(buzzerPin);
    delay(300);
  }
  delay(100);
}
```

The display is updated inside the loop function. This is also where you'll check that the battery voltage hasn't dropped below the warning voltage and check for key presses to toggle the battery monitor's mute mode.

This loop function makes use of a number of other functions further down in the file. The first of these is displayVoltage.

```
void displayVoltage()
{
  lcd.setCursor(8, 0);
❶ lcd.print("        ");
  lcd.setCursor(8, 0);
❷ lcd.print(readVoltage());
  lcd.setCursor(14, 0);
  lcd.print("V");
}
```

This function starts at column 8 and overwrites the eight character positions on the top line by printing eight spaces ❶. It then moves the cursor back to column 8 and writes the battery voltage in that gap ❷ before writing the *V* character at the end of the line.

The displayVoltage function makes use of the readVoltage function to convert the raw reading from the Arduino's analog input into a voltage.

```
float readVoltage()
{
  int raw = analogRead(voltagePin);
  float vout = (float(raw) / 1023.0) * 5.0;
  float vin = (vout * k);
  return vin;
}
```

Readings from an Arduino analog pin give a result between 0 and 1,023, where 0 means 0V and 1,023 means 5V. So, the value of vout in readVoltage is the output voltage of the potential divider—that is, the reduced voltage. You need to work backward to calculate the original battery voltage vin, then return this value to be displayed.

The final function in the sketch displays the bar graph showing how much power is left in the battery and, if the battery monitor is in mute mode, the MUTE notification.

```
void displayBar()
{
  float v = readVoltage();
  float range = maxV - minV;
  float fullness = (v - minV) / range;

  int numBars = fullness * 16;
  lcd.setCursor(0, 1);
  for (int i = 0; i < 16; i++)
  {
    if (numBars > i)
    {
      lcd.print("*");
    }
    else
    {
      lcd.print(" ");
    }
  }
  if (mute)
  {
   lcd.setCursor(12, 1);
   lcd.print("MUTE");
  }
}
```

The displayBar function steps through each of the 16 character positions of the second row of the display and then displays either a * or a space character, depending on the measure of fullness of the battery.

USING THE BATTERY MONITOR

As soon as you connect the battery monitor to the battery, the LCD should light up and show a readout of the battery voltage on the top row of the display. The second row of the display will show a number of * characters to indicate the juice remaining in the battery. Also, if you press any of the button switches below the display to disable the buzzer, the message MUTE should toggle on and off.

If your display appears blank or difficult to read, then you may need to adjust the contrast. Just use a small, flat-headed screwdriver to turn the small variable resistor at the top right of the LCD shield (Figure 3-11).

Now that you have the basics of power generation and lighting sorted out, turn your attention to detecting zombies. You'll find out how to know they're coming in Chapter 4.

4

ZOMBIE ALARMS

 Movies tell us that zombies can't move around without groaning. They're also clumsy and liable to crash into things. However, there's still the possibility that they will catch you unaware. After all, you have to sleep sometime. So, one of the first uses of your newly generated electricity should be to make some zombie alarms (Figure 4-1).

This chapter has two zombie detector projects: a decidedly low-tech trip wire alarm and a more sophisticated passive infrared (PIR) proximity alarm.

FIGURE 4-1: ZOMBIE DETECTION

PROJECT 5: TRIP WIRE ALARM

Zombies will keep finding their way into your compound, either because
they're attracted to the smell and noise or just through aimless wandering.
You need a way to detect them so that you can grab a baseball bat or ax and
head off to do battle at the breach in your defenses.

Alternatively, you may decide to create a "killing field" into which unsus-
pecting zombies (is there another kind?) will wander, ready for swift dis-
patching. Either way, you'll need to be alerted to their presence, and a trip
wire is a good way to make sure that happens.

Zombies are notorious for dragging their feet. They also frequently
fail to look where they're going, since they're mostly guided by the smell
of human flesh. So, even a trip wire that wouldn't fool the most clumsy
and shortsighted of humans will work just fine on a zombie (Figure 4-2).
This alarm uses parts that are easily scavenged to sound a car horn when
triggered.

FIGURE 4-2: A TRIP WIRE ALARM

WHAT YOU WILL NEED

To make this project, you'll need the following items. (The microswitch can also be obtained from the door safety interlock of a microwave.)

ITEMS	NOTES	SOURCE
☐ string	Long enough to stretch across the gap where you want to detect zombies	Hardware store
☐ Nails or screws	To fix the trip wire and microswitch	Hardware store, scavenge
☐ microswitch		Fry's (2314449), microwave oven
☐ Double-core bell wire or speaker cable	To connect the microswitch to the battery and car horn	Hardware store, scavenge
☐ car horn	The louder the better. Even zombies can look surprised when a car horn blares a few feet from their head, and few things are funnier than a surprised-looking zombie.	Auto parts store, scavenge
☐ 12v battery	This can be a car battery, but a smaller battery will also be just fine.	Auto parts store, scavenge

You could scavenge most of these parts easily, and you probably won't have enough electricity to run a microwave, so you may as well strip it down for the microswitch. Of course, if you're practicing before the zombies have completely taken over, it would clearly be a terrible waste to destroy a microwave oven just for a $2 switch; in that case, only use a microwave that is already dead. It's dangerous to keep a zombie microwave around, anyway.

The 12V battery I used is a small, sealed lead-acid battery. These are effectively miniature car batteries. But if you have a car battery all set up from Chapter 2, then you will probably just want to use that one.

WARNING The following procedure describes how to disassemble a microwave oven. You must only do this with the oven unplugged from the AC outlet. This procedure will render the oven at best nonfunctional and at worst very dangerous, so the microwave must be scrapped after this. Do not attempt to use the damaged microwave as a radiation weapon against the zombies (but using it as a blunt object is okay).

CONSTRUCTION

In a postapocalyptic world, low-tech traps, like this one, will usually be the most reliable. The hardest part of the build is probably extracting the microswitch from a microwave oven, and that's where I'll start. Of course, if you plan ahead, you can also just buy a switch; skip to "Step 2: Identify the Microswitch Terminals" on page 68 if you already have your switch prepared.

As with most of the projects in this book, it is a good idea to get them working safely on your workbench before deploying them in an active zombie area. It's very difficult to concentrate on your soldering when a groaning heap of rotting flesh is bearing down on you.

STEP 1: OBTAIN A MICROSWITCH

All microwave ovens are slightly different, so you'll need to adapt these instructions to your particular microwave. The basic principle is to keep taking out screws and removing parts of the microwave until you get to the door switch. Most microwave ovens have a U-shaped outer case that, once removed, gives you a clear view of the oven's inner workings (Figure 4-3). The microswitch next to the door latch is then easily accessible.

In Figure 4-3, the microwave door is to the right, the back of the area with all the control buttons and knobs is at the top right, and the inside of the door latch is close to the circled area.

FIGURE 4-3: INSIDE THE MICROWAVE OVEN

The microswitch you are trying to find will have clips attached to its terminals (Figure 4-4). You can pull those clips off, or you can simply cut the wires down to the clip to make the microswitch easier to remove.

After you remove the microswitch, if the attached wires are long enough, you can just leave them in place and cut them. Otherwise, desolder the wires and attach new wires of the correct length. The microwave also contains a lot of other useful lengths of wire, particularly those with spade terminals attached.

Unfortunately, not much more of the oven is going to be of use in this project, though the remainder of the microwave oven makes an extremely effective zombie head crusher if dropped from a height into the groaning host of undead. Just attach a rope to it first so you can reel it back and use it multiple times. There's an apocalypse on, so it's even more important to recycle.

STEP 2: IDENTIFY THE MICROSWITCH TERMINALS

The microswitch will have three connections (Figure 4-4). If you look closely, you'll see that they are marked COM (common), NC (normally closed), and NO (normally open). When pushed toward the left, the long lever pushes in the little button on the side of the switch.

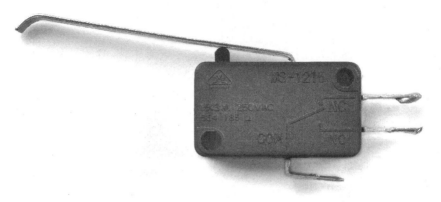

FIGURE 4–4: A MICROSWITCH

You'll always use the common connection when you add one of these switches to a circuit. The other terminal you connect to will depend on whether you want the switch to cause something to happen when it's activated or when it's released. The normally open terminal of the switch is left open when the switch is not activated, meaning it has no connection to the common terminal until the button is pressed. The normally closed terminal works the other way around. You want your alarm to go off when a zombie hits the trip wire, thus activating the switch, so this project uses the NO connection.

STEP 3: PREVIEW THE ELECTRONIC CIRCUIT

Now let's have a look at how that little switch will become a key part of your advance zombie-warning system.

The schematic of the trip wire alarm (Figure 4-5) shows an electronic circuit that is about as simple as one can get. If one of the car horn's terminals is explicitly marked positive, you'll connect it to the positive terminal of the battery; otherwise, it doesn't matter which side of the horn you attach there. The battery's negative terminal is connected to the microswitch's COM terminal, and the microswitch's NO connection completes the circuit back to the car horn.

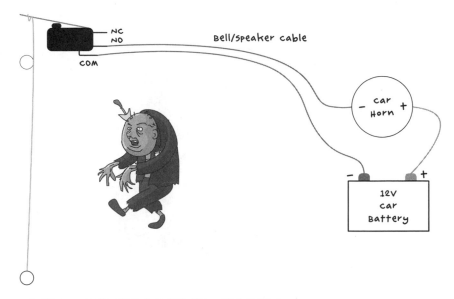

FIGURE 4-5: SCHEMATIC FOR THE TRIP WIRE ALARM

car horns normally require a full 12V before they'll make much noise. Naturally, a car battery is well suited for this project. See chapter 2 for more information on using batteries.

One big advantage of this alarm is that it doesn't draw any current at all from the battery until the alarm is triggered. This means that your battery will provide effective zombie protection for a long while.

STEP 4: PREPARE THE WIRES

You're going to need two lengths of cable: a longer length of double-core cable leading to the switch and a short single length (perhaps 6 inches, or 15 cm) of wire to connect the positive terminals of the battery and horn.

Strip and trim the ends of all the wires. If you need help doing this, see "Stripping Wires" on page 227.

STEP 5: CONNECT THE BATTERY AND HORN

If you're using a small 12V battery with solder tabs, solder the short single length of wire between the positive terminal of the battery and the positive terminal of the horn (Figure 4-6). If, on the other hand, you're using a car battery, then don't try to solder directly to the terminal. Instead, use an alligator clip, as you did with many of the leads in Chapter 2 (see, for example, Figure 2-10 on page 32).

FIGURE 4-6: CONNECTING THE HORN AND BATTERY

Again, if the horn doesn't have a terminal marked with a +, then it doesn't matter which terminal of the horn you connect the battery to.

STEP 6: CONNECT THE SWITCH

Solder the two wires of one end of the long length of double-core cable to the COM and NO terminals of the switch (Figure 4-7). It doesn't matter which wire goes to which switch terminal, as long as you use COM and NO. In fact, sometimes there won't be an NC terminal on the microswitch at all.

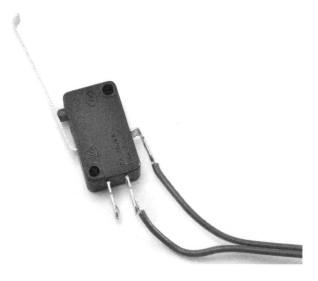

FIGURE 4-7: SOLDERING WIRES TO THE SWITCH

Now connect the other ends of the double-core cable to the unused terminal of the horn and the negative terminal of the battery (Figure 4-8).

FIGURE 4-8: THE FINISHED WIRING

With all the wiring complete, you'll find that when you activate the switch by moving the lever, the horn will sound. Car horns are really loud, so it's best not to test this out in an enclosed space—and warn your fellow survivors before you try it!

USING THE TRIP WIRE ALARM

Clearly, you need to choose your moment to deploy this project so that you don't become a zombie yourself before you finish setting it up. When you've identified the opening that you want to protect, fix a screw or nail (or find some other way of securing one end of the string) about 6 inches (15 cm) off the ground. This will allow the switch to be triggered both by foot-dragging zombies and more athletic zombies that may simply tread on the string.

Fix the microswitch on the opposite side of the passageway you're protecting and at the same height as the anchor for the string. Most microswitches have holes that make them easy to affix with small screws. If not, you can glue the switch in place with epoxy-based glue or a hot glue gun.

Position the switch so that the lever is on the other side from the space being protected. Tie the string around the top end of the lever (Figure 4-9).

FIGURE 4-9: THE SWITCH FIXED IN PLACE

Don't make your string too taut. After all, you want it to detach without pulling the microswitch off the wall when a zombie walks through it. Tying a bow at one end or the other is a good idea.

Although I used a car horn, you could use anything that makes a noise and operates from 12V. Or, if you'd prefer a silent alarm, you could use a 12V car light bulb in place of the car horn.

Neither approach is very sophisticated, however, so in the next project, you'll level up your zombie alarm with something a little more high-tech.

PROJECT 6: PIR ZOMBIE DETECTOR

The second zombie detector project in this book uses a *passive infrared (PIR) detector*. These detectors are the same type used in intruder alarms—they sense movement of heat—and I guess few things are more intrusive than a group of zombies intent on eating you.

You can of course just buy (or scavenge) an intruder alarm that uses PIR sensors, rather than make this project from scratch, but I thought it would be more fun to make something that uses an Arduino. In fact, if you just add the extra components needed for the PIR alarm to "Project 4: Battery Monitor" on page 53, the same Arduino can both monitor your battery and alert you to a zombie attack, using the same buzzer and display.

When a zombie triggers the PIR sensor, the LCD display will show the message *ZOMBIES!!* (Figure 4-10). Since the last thing you want to do while fending off a zombie is attract more zombies, this project also allows you to silence the alarm by pressing any button on the LCD shield.

FIGURE 4–10: PIR ZOMBIE DETECTOR

WHAT YOU WILL NEED

To make this PIR alarm, you'll need the following parts. If you've already made the battery monitor of "Project 4: Battery Monitor" on page 53, you'll already have the Arduino, screwshield, and alligator clips.

ITEMS	NOTES	SOURCE
☐ Arduino	Arduino Uno R3	Adafruit, Fry's (7224833), Sparkfun
☐ Arduino screwshield	screwshield	Adafruit (196)
☐ PIR module		Adafruit (189), Fry's (6726705), security store
☐ Small alligator clip leads		Auto parts store
☐ 3-core cable wire	Long enough to reach the PIR sensor	Scavenge
☐ Terminal block	3-way, 2A terminal strip	Auto parts store, electrical store

CONSTRUCTION

This is another project that can be assembled without any soldering, and my
instructions assume that you're building on top of Project 4. If you haven't
already built Project 4, then you'll need to build a slightly modified version
of that project first, as the hardware for the PIR zombie detector is mostly
the same.

STEP 1: BUILD THE SCREWSHIELD

Flip to Project 4, "Construction" on page 55 and follow Steps 1 to 3. In
Step 1, download the sketch (Arduino's word for program) *Project_06_PIR
_Alarm* from *http://www.nostarch.com/zombies/* and use that in place of the
sketch for Project 4. Also, when it comes to Step 3, you don't need to include
the two resistors unless you also want to monitor the battery voltage.

STEP 2: MAKE A LEAD FOR THE PIR SENSOR

There's little point in making a zombie detector that detects zombies only
after they're already in the same room as you. Chances are you'll have
already thoroughly detected them if they get that far. Therefore, you need
to attach a long lead to the PIR detector so that it can monitor the corridor,
porch, or other area outside your living space.

The PIR detector has three leads: two that supply power and one output that indicates that motion has been detected. This means you'll need a
three-wire lead. You could find some wire from an intruder alarm, or you
could use three of the wires in a telephone extension lead. Pretty much any
lead with three or more wires in it will be just fine.

You could either solder the ends of this lead to the lead that comes with
the PIR sensor or use a terminal block, as I have (Figure 4-11).

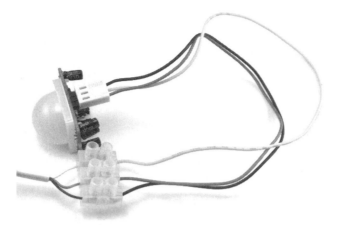

FIGURE 4—11: PIR LEAD AND TERMINAL BLOCK

I harvested my three-lead cable from a telephone extension lead. The cable contained four solid-core insulated wires. These wires were color coded, so I used blue for GND (ground), orange for 5V, and stripy white for the output; I left the final wire unused. The lead was about 30 feet (9 m) long, which worked fine for the sensor. You can probably use longer leads, but try it out first before you lay all the cabling.

STEP 3: CONNECT THE PIR TO THE SCREWSHIELD

Now that you've extended your sensor wires to a useful length, attach the wires to the Arduino screwshield (Figure 4-12; note that the two resistors from Project 4 are shown at the bottom left).

FIGURE 4—12: CONNECTING THE PIR LEAD TO THE ARDUINO SCREWSHIELD

If you look at the back of the PIR sensor, you'll see that the three pins are labeled GND, OUT, and +5V. Connect GND on the PIR sensor to one of the GND connections on the screwshield; it doesn't matter which one. Then, connect +5V on the PIR sensor to the 5V connection on the screwshield. Finally, connect OUT on the PIR sensor to D2 on the screwshield.

SOFTWARE

If you just want to make this project on its own, without any of the earlier Arduino-based projects, then use the sketch *Project_06_PIR_Alarm*. On the other hand, if you've made one or more of the other Arduino projects and wish to include them, then use the sketch *All_Sensors* and change the constants at the top to select the projects that you have made.

The first few lines of the *All_Sensors* sketch are shown below:

```
/*
Any projects that you want to exclude from this program should have a
value of "false". That way, you will not get any false alarms because
of missing hardware.
*/
const boolean project4 = true; // Battery Monitor
const boolean project6 = true; // PIR Alarm
const boolean project10 = false; // Door Monitor
const boolean project11 = false; // Fire Alarm
const boolean project12 = false; // Temperature Monitor
```

In this case, only the battery monitor (Project 4) and PIR alarm (Project 6) are enabled. If you've made more of the projects, then change the value next to those projects from false to true. If you are working your way through this book in order, then the program should look as shown.

All the source code for this book is available from *http://www.nostarch.com/zombies/*. See Appendix C for instructions on installing the programs.

The PIR detector code follows the same pattern as Project 4, so for more information on how the program as a whole works, please refer to "Software" on page 57. Here, I'll just describe the code specific to this project.

The first change to the earlier code is the addition of a new constant for the PIR's OUT pin. I added the pirPIN constant on the line after the switchPin constant.

```
const int pirPin = 2;
```

I set `pirPin` to 2 because the output of the PIR sensor will be connected to pin 2 on the Arduino. The next addition to the sketch occurs in the setup function, where that same pin 2 is set to be an input.

```
pinMode(pirPin, INPUT);
```

Although pins on an Arduino default to inputs unless specified as outputs, declaring the pin to be an input makes the code easier to follow.

The loop function now needs to check the sensor, so I added a call to the function `checkPIR`.

```
checkPIR();
```

This new function, `checkPIR`, will, as the name suggests, check the PIR sensor and take the appropriate action if the sensor is triggered. The function is defined right at the end of the sketch.

```
void checkPIR()
{
  if (digitalRead(pirPin))
  {
    alarm("ZOMBIES!!");
  }
}
```

The `checkPIR` function makes a `digitalRead` of the `pirPin` to decide whether the output from the PIR detector is `HIGH` or `LOW`. If movement has been detected, then the `alarm` function is used to display an appropriate message. For more information on using the inputs and outputs of an Arduino, see Appendix C.

USING THE PIR ZOMBIE DETECTOR

The project works well in combination with the battery monitor, as you can just run both off the same battery. But whether you combine the two projects or not, be mindful of your wires when you deploy the PIR detector around your base of operations. If you're using cable that contains solid-core wires, then affix the cable to the wall at regular intervals along the cable length. Solid-core wires don't take kindly to being repeatedly flexed.

SCAVENGED PIR SENSORS

The Adafruit PIR module used in this project is designed to work with microcontroller modules like the Arduino. But following an apocalypse, you may find it easier to obtain the type of regular PIR sensor intended for use with a security system, such as the unbranded unit obtained from eBay for a couple of dollars, shown opened up in Figure 4-13.

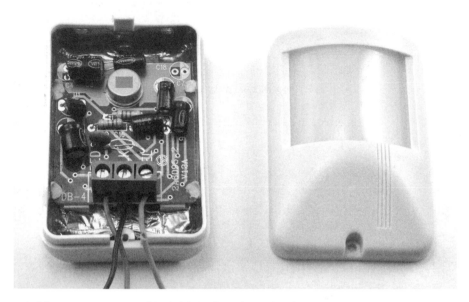

FIGURE 4–13: A PIR MODULE INTENDED FOR INTRUDER ALARMS

This sensor won't operate at 5V but rather requires a power supply of 12V. The sensor has a logic level output that will rise to 3.6V, which is enough to register as HIGH, just like the Adafruit module. The only difference in wiring is to connect this sensor's red wire to the Arduino's V_{in} rather than to 5V.

Be aware that other sensors may look like this one but have a different output voltage. Some (with an open collector output) require a pull-up resistor (of, say, 1 kΩ) between the output and 5V on the Arduino. If the output of the sensor does not give a useful voltage when you wave your hand in front of it, then it almost certainly needs a pull-up resistor.

Other types of PIR sensors, especially those intended to control lighting, have a relay output. This output works just like a switch, closing when movement is detected. The schematics show how to connect three types of PIR modules to the Arduino (Figure 4-14).

Wherever possible, choose a device that you have documentation for so you don't have to guess how its output works and how to wire it up.

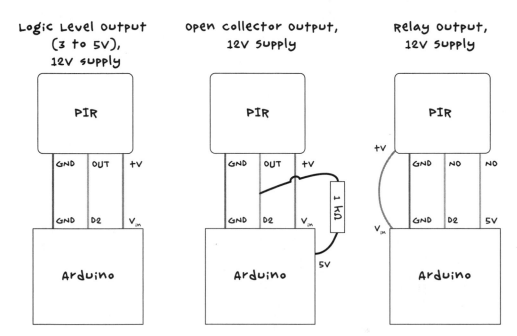

FIGURE 4–14: CONNECTING DIFFERENT TYPES OF PIR MODULE TO THE ARDUINO

The next chapter advances from automatic zombie detection to walk through a number of surveillance projects that will allow you to see what is going on before it trips over your doorstep. You'll be able to spot the zombies remotely using webcams.

5

SURVEILLANCE AND RASPBERRY PI

Now that you can detect zombies, it's also a good idea to monitor their movements. But don't risk joining the undead ranks by following them around! Watch them safely from inside your base, and you'll keep your brain intact. This chapter shows you how to make a surveillance camera setup with USB and wireless webcams, using a Raspberry Pi single-board computer to minimize your energy usage (see Figure 5-1).

Both projects in this chapter require you to download software, so you'd be well advised to think ahead and get your system set up before disaster strikes.

FIGURE 5–1: ZOMBIE SMILING AND WAVING AT A WEBCAM

THE RASPBERRY PI

You could get these projects working with a regular laptop or desktop computer, but those devices take a fair bit of power. A laptop typically consumes 20W to 60W, and a desktop draws even more. Also, you'd need an AC inverter. Laptop power supplies provide low-voltage DC, but generally that voltage is still higher than 12V, so powering directly from a 12V battery wouldn't be an option.

Besides, if you have to shift bases because the zombie population density has gotten too high, do you really want to risk being weighed down by a giant desktop tower?

The Raspberry Pi, on the other hand, is a tiny Linux computer on a single board about the size of a credit card, and it uses less than 3W of power. A Raspberry Pi Model B+ is used in this project and throughout this book (Figure 5-2). If you happen to have an older Raspberry Pi Model B or a newer Raspberry Pi 2, they should also work just fine. In fact, the extra power of the Pi 2 should make the webcam browser page perform noticeably quicker. Models A and A+ are not ideal, as they are less powerful and have less memory than the other models.

FIGURE 5-2: A RASPBERRY PI MODEL B+

The Raspberry Pi can run simple Python scripts, and you can link it to external hardware, too. For example, in "Project 7: Monitor Zombies with a USB Webcam" on page 87, when the webcam detects movement, an LED will turn from green to red using the Raspberry Pi's GPIO (general purpose input and output) connector. The GPIO connector is the double row of pins down one side of the board (Figure 5-2).

THE RASPBERRY PI SYSTEM

A complete Raspberry Pi system includes a USB keyboard, a mouse, and a small HDMI (High-Definition Multimedia Interface) monitor (Figure 5-3).

The keyboard and mouse are standard items that you can buy anywhere. For a constant visual on your zombie foes, you'll need something to watch the video feed on, and you could just connect a normal TV or monitor to the Raspberry Pi. However, to save even more power, this project uses a 12V DC monitor with a 7-inch (180 mm) display. At worst, this might double the power consumption to a peak of 6W.

FIGURE 5-3: A RASPBERRY PI SYSTEM

WHAT YOU WILL NEED

To use this Raspberry Pi system with a 12V battery as this book describes, you'll need the following items.

ITEM	NOTES	SOURCE
☐ Raspberry Pi	Model B+ or Pi 2 with NOOBS micro SD card	Adafruit (2358), Fry's (8258726)
☐ Small HDMI monitor	12V HDMI monitor. Suggested device has 800×480 pixel resolution.	Adafruit (1934), eBay
☐ Keyboard and mouse	Standard USB keyboard and mouse	computer store, online
☐ HDMI cable	As short as possible	computer store, online
☐ 12V to USB adapter	Minimum current of 1 A	Auto parts store, computer store
☐ vehicle to 2.1 mm jack adapter		Auto parts store
☐ Powered USB hub	Needed only if you have a Raspberry Pi Model B	computer store, online

If you're using a Model B Raspberry Pi that has only two USB sockets, then you'll need a powered USB hub or a wireless keyboard and mouse combo that uses a single USB adapter. Otherwise, the keyboard and mouse will occupy both of the Model B's USB ports, and you won't be able to plug in the webcam needed in the next project.

POWERING THE SYSTEM

The Raspberry Pi is powered from a micro USB socket, so you can use a 12V-to-USB power adapter when powering it from a 12V battery. The monitor I suggest has a separate driver board that powers the display and connects it to the Raspberry Pi; that's the printed circuit board (PCB) in the middle of Figure 5-3. This driver board has a 2.1 mm DC power socket.

A combined cigarette lighter and USB socket adapter (such as in Figure 5-4) is a great way to power this whole system from batteries. If you haven't already done so, you'll need to replace the cigarette plug with a pair of alligator clips to attach the adapter to the battery. Refer to Chapter 3 for instructions on how to connect your 12V battery to low-voltage devices.

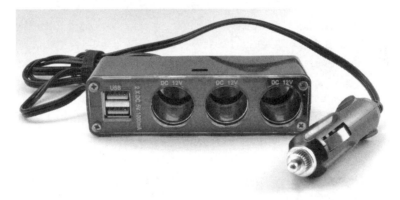

FIGURE 5-4: COMBINED USB AND 12V DC POWER ADAPTER

With the power adapter setup of Figure 5-4, you can power your Raspberry Pi from a normal micro USB lead and, in "Project 8: A Wireless Zombie Surveillance System" on page 96, power the Wi-Fi webcam and router with a DC jack-to-cigarette lighter adapter. Check the voltages used by your router and Wi-Fi webcam, but they're quite likely 12V DC, which is very handy if you've stockpiled car batteries for the apocalypse already.

WARNING   Be careful when handling the display, especially if the display has a metal back. The exposed underside of the driver board can easily short against the metal, damaging the board.

To connect the driver board to a car battery from your stockpile, just make a lead with a 2.1 mm jack on one end and alligator clips on the other. However, if your battery is overloaded with alligator clips, you may want to attach a multiple cigarette lighter socket adapter to it instead. Then you can plug various appliances into the adapter with cigarette lighter plugs, as described in "Cigarette Lighter Sockets" on page 46.

INSTALLING RASPBIAN

The Raspberry Pi computer doesn't have a hard disk. Instead, the Raspberry Pi 2 and Model B+ stores its operating system, programs, and data on a micro SD card. Older Raspberry Pi models store that information on a regular SD card. There won't be an Internet after the zombie apocalypse, so get a micro SD card preloaded with an operating system (OS)—you won't be able to download it. In fact, a Raspberry Pi with a preloaded SD card usually doesn't cost much more than the Raspberry Pi on its own, so I recommend just buying the preloaded card with your Raspberry Pi. If you do want to add an OS to a blank SD card yourself, visit *http://www.raspberrypi.org/help/noobs-setup/* and follow the directions there before the Internet ceases to exist.

Whether you buy a preloaded micro SD card or add the software yourself, this book assumes you're using a micro SD card with the Raspberry Pi Foundation's NOOBS (New Out Of the Box Software) installer. Once you have one, fit the micro SD card into the Raspberry Pi; plug in the keyboard, mouse, and monitor; and power everything up.

NOTE The monitor I suggest for this project should detect the Raspberry Pi through the HDMI cable, and the Pi should automatically detect the screen resolution. If the Pi doesn't detect the screen resolution, then visit the Raspberry Pi's documentation page (http://www.raspberrypi.org/documentation/), go to the configuration section, and read config.txt to learn how to configure your Raspberry Pi. Print the instructions and keep them with this book so you're ready when the apocalypse ends the Internet as we know it.

When you boot the Raspberry Pi with NOOBS, you'll be offered your choice of operating system. This book uses Raspbian, so select the checkbox next to Raspbian and then click **Install**. The installation will take a while, so watch your PIR zombie detector or double-check your battery stockpile while you wait. Once the installer finishes, you're ready to move on.

The Raspbian distribution comes with a pretty comprehensive set of software, but at the time of writing, one thing it lacks is a decent browser that will work with a webcam. I favor Chromium, a derivative of Google Chrome that works well without hogging so many of the Raspberry Pi's resources that the Pi becomes too zombie-like for comfort. As with most free software, you'll need to download Chromium from the Internet.

I apologize if it's too late, but if it's not, then connect the Raspberry Pi to your preapocalyptic home modem or router with an Ethernet cable. Then, to install Chromium, click the **LX Terminal** icon on the Raspberry Pi desktop. A terminal window should open, and at first, you should just see a flashing cursor and a command prompt like this:

```
$
```

Anytime you need to enter commands for a project in this book, I'll also show the dollar command prompt on the left, which you don't need to type. Now, enter the following commands:

```
$ sudo apt-get update
$ sudo apt-get install chromium
```

The sudo (short for *substitute user do*) command allows you to execute administrative commands. Prepend it to commands that need administrative access, such as commands that install new software as we're doing now.

The apt-get package management software on Debian-based Linux distributions such as Raspbian is used to manage and install software. The update command used with apt-get tells your system to update its cached list of available software from Internet software repositories. The apt-get install command tells apt-get to search for and install the latest version of the package supplied as the final argument, which in this case is Chromium.

With Chromium in place, you are ready to build your surveillance system. Now let's monitor some zombies!

PROJECT 7: MONITOR ZOMBIES WITH A USB WEBCAM

This project uses a low-cost USB webcam with a long lead attached to the Raspberry Pi. The maximum usable length of a USB 2 lead is 96 feet (30 m), so that's the farthest away from the Raspberry Pi that your camera can be.

You can see most of the setup in Figure 5-5, though the webcam is just out of view on the left; I've shown it in the inset photo. One of the benefits of building a surveillance system for yourself rather than simply using an off-the-shelf closed-circuit television (CCTV) system is that because the software is completely under your control, you can customize it however you want.

The webcam is controlled by a short Python program that monitors the images being captured for changes. When movement is detected on the screen, the program uses the Raspberry Pi's GPIO pins to turn an RGB (red-green-blue) LED from green to red. You can cancel the alarm by pressing the spacebar on the keyboard, which will turn the LED green again.

FIGURE 5-5: ZOMBIE WEBCAM AND MOVEMENT ALARM

The advantage of this project over "Project 6: PIR Zombie Detector" on page 72, which uses a PIR sensor, is that now, if the alarm is triggered, you can take a good look at the zombies that are about to attack you.

WHAT YOU WILL NEED

To set up this USB webcam, you'll need the Raspberry Pi setup described in "The Raspberry Pi" on page 82 and the additional items described here.

ITEM	NOTES	SOURCE
☐ USB webcam	See http://elinux.org/RPi_USB_Webcams/ for compatible webcams.	computer store
☐ USB extension lead	Length to suit your compound (less than 100 feet [30 m])	computer store
☐ Raspberry squid	contains the RGB LED	Amazon, http://www.monkmakes.com/

Not every USB webcam is compatible with the Raspberry Pi, so check *http://elinux.org/RPi_USB_Webcams* for a list of cameras known to work with the Raspberry Pi. I used an HP 2300 Webcam.

NOTE *The Raspberry Pi camera module is a high-resolution camera that plugs directly into a special connector on the Raspberry Pi. The module is great if you're making a Raspberry Pi camera, but it's not much use in a situation like this, where you want the camera to be some distance from the Raspberry Pi.*

The Raspberry Squid is a handy little accessory built just for the Raspberry Pi. It has an RGB LED with built-in current-limiting resistors that allow you to connect it directly to the Raspberry Pi's GPIO pins. Its design is open source, and you can find details of how to build your own here: *https://github.com/simonmonk/squid/*. You can also buy a ready-made Squid; see *http://www.monkmakes.com/* for details.

CONSTRUCTION

After completing the setup in "The Raspberry Pi System" on page 83, to build this project you just need to attach the Raspberry Squid to the GPIO connector of the Raspberry Pi, plug in the USB webcam, supply 12V to the monitor, and supply 5V to the Raspberry Pi (see Figure 5-6).

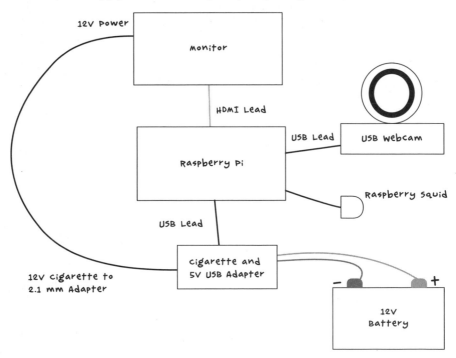

FIGURE 5-6: SCHEMATIC FOR THE SURVEILLANCE SYSTEM

STEP 1: ATTACH THE RASPBERRY SQUID

By controlling the three outputs of the Raspberry Squid, you can make the LED display any color. However, this project won't use the accessory's full potential since this surveillance setup needs to display only red and green.

To help identify the GPIO pins, you can use a GPIO pin identification template. There are many of these available from suppliers like Adafruit, including the Raspberry Leaf, which is included if you buy a ready-made Raspberry Squid. Place this template over the GPIO connectors so that you can tell which pin is which. Then connect the Raspberry Squid to the GPIO connector (Figure 5-7).

FIGURE 5-7: CONNECTING THE RASPBERRY SQUID TO THE GPIO CONNECTOR

The black lead of the Raspberry Squid goes to one of the GND pins on the Raspberry Pi. In the orientation shown (Figure 5-7), this is the third pin down on the right. The red lead of the Raspberry Squid goes to pin 18 on the Raspberry Pi, and the green lead of the Squid goes to pin 23 of the Pi. Since you won't need the blue color, you can leave the blue lead of the Raspberry Squid unattached, but if you prefer to keep the leads tidy, just attach the blue lead to any one of the other GND pins of the GPIO header.

STEP 2: INSTALL THE USB WEBCAM

If you already have a USB webcam, then see if it works with the Raspberry Pi before you get another one. First, check whether the Raspberry Pi can detect your webcam as a USB device by entering the command **lsusb** in LXTerminal both before and after plugging the webcam into the Pi, without the USB extension lead.

```
$ lsusb
Bus 001 Device 002: ID 0424:9514 Standard Microsystems Corp.
Bus 001 Device 001: ID 1d6b:0002 Linux Foundation 2.0 root hub
Bus 001 Device 003: ID 0424:ec00 Standard Microsystems Corp.
Bus 001 Device 004: ID 03f0:e207 Hewlett-Packard
Bus 001 Device 006: ID 04d9:1603 Holtek Semiconductor, Inc. Keyboard
Bus 001 Device 005: ID 1c4f:0034 SiGma Micro
```

If an extra entry appears after you run the command with the webcam plugged in, then that entry should be your webcam. In the list I've shown, my Hewlett-Packard webcam is the fourth entry from the top.

If your webcam does not appear in the list, then try unplugging it, plugging it back in, and running the lsusb command again. If that doesn't work, try a reboot of the Raspberry Pi.

Unfortunately, being recognized as a USB device is still no guarantee that a webcam will work with the Raspberry Pi. You'll find out for certain when you run the program. You may also find that your webcam works only if it's plugged into a powered hub. If you have an older model of Raspberry Pi, you may find instead that the whole board resets when you plug the webcam into the USB port. If this is the case for your board, then plug the webcam in while the Raspberry Pi is powered off.

STEP 3: INSTALL THE SOFTWARE

Connect the Raspberry Pi to your network with an Ethernet cable, make sure the Internet is up and running, and download the Raspberry Pi programs for the projects in this book. From your browser on the Pi, you can head to *http://www.nostarch.com/zombies/*, click the link to GitHub, and download the *Raspberry Pi* directory. For this project, you'll use the code in the *usb_webcam* directory. But the easiest way to get the software onto your Raspberry Pi is to clone the GitHub repository directly onto your Raspberry Pi, as I describe in "Fetching Source Code from GitHub" on page 92.

The Python program *monitor.py* is pretty brief, considering what it does, and I'll walk you through it here. I won't, however, cover Python itself beyond the context of the projects that use it. If you are new to Python, you might take a look another of my books, *Programming the Raspberry Pi: Getting Started with Python* (McGraw-Hill, 2013).

The program begins by importing the various Python modules that it needs. These libraries of existing code are all included in the Raspbian distribution, so you shouldn't need to install them separately.

```
import sys
import time
import pygame
import pygame.camera
import RPi.GPIO as GPIO
```

The sys and time modules have general utilities for accessing the operating system and the ability to tell the program to sleep as a way of delaying its activity for a period of time. The pygame module contains the Pygame graphical games library, which includes a camera interface. To control the LED, the program needs access to the GPIO system, and this is provided by the RPi.GPIO library.

Next, the program defines some constants that it will use. You could change these if you wanted to use the camera at a different resolution or make the default size of the window larger.

```
camera_res = (320, 240)
window_size = (640, 480)
red_pin = 18
green_pin = 23
```

The parameters in parentheses after the `camera_res` and `window_res` constants are the width and height respectively (in pixels). After the constants, the Pygame system (used to display the camera images) and the camera itself are initialized, along with the GPIO ports that you'll use to control the Raspberry Squid:

```
❶ pygame.init()
  pygame.camera.init()

  # initialize GPIO
❷ GPIO.setmode(GPIO.BCM)
  GPIO.setup(red_pin, GPIO.OUT)
  GPIO.setup(green_pin, GPIO.OUT)

❸ screen = pygame.display.set_mode(window_size, 0)

  #Find, open, and start the low-res camera.
❹ cam_list = pygame.camera.list_cameras()
  webcam = pygame.camera.Camera(cam_list[0], camera_res)
  webcam.start()
❺ old_image = False
```

The first two lines of initialization code ❶ handle Pygame and the camera, while the next three lines ❷ initialize those GPIO ports. The screen is then initialized ❸ to the size of the window specified in `window_size`. The final cluster of lines ❹ first finds all the cameras connected to the Raspberry Pi and then creates a link to the first one (`webcam`). It then starts running the webcam. The final line ❺ defines a variable called `old_image`, which is used to detect movement by spotting changes in successive frames from the webcam.

After initialization, the first function this program defines is called `check_for_movement`.

```
def check_for_movement(old_image, new_image):
    global c
    diff_image = pygame.PixelArray(new_image)
      .compare(pygame.PixelArray(old_image), distance=0.5,
      weights=(0.299, 0.587, 0.114))

    ys = range(0, camera_res[1] / 20)
    for x in range(0, camera_res[0] / 20):
        for y in ys:
            if diff_image[x*20, y*20] > 0:
                return True
    return False
```

As the name suggests, `check_for_movement` takes two images, the previous frame (`old_image`) and the latest frame (`new_image`), and compares them. The `distance` parameter to `compare` is the "distance" between the color of the pixel in one image and the color of that same pixel in the other image. The `weights` parameter is not explained in the `pygame` documentation, and the values used here are taken in faith from an example in the `pygame` documentation for `PixelArray` (*http://www.pygame.org/docs/ref/pixelarray.html*).

The comparison results in a new image called `diff_image` that only has white pixels where a difference was found between the pixels in the two images.

To decide whether movement has occurred, the program should really go through every pixel in the `diff_image`. But any largish movement will result in lots of pixels changing, and zombies are big, so the code speeds things up by only sampling 1 pixel in 20.

The next two functions set the LED of the Raspberry Squid to red or green.

```
def led_red():
    GPIO.output(red_pin, True)
    GPIO.output(green_pin, False)

def led_green():
    GPIO.output(red_pin, False)
    GPIO.output(green_pin, True)
```

The Raspberry Squid is just an RGB LED, and as with most RGB LEDs, you can select the color it glows by outputting certain combinations of high (`True`) and low (`False`) on the GPIO pins the LED is connected to. In this case, you want red and green, so the code just sets the appropriate pin to `True` and the other to `False`. The blue takes no part in this project, so you don't have to deal with it in the code.

Finally, we come to the main loop of the program, where the new image is fetched and scaled so it's ready to display in the window.

```
count = 0
led_green()
while True:
    count = count + 1
    new_image = webcam.get_image()
    # Set old_image the first time around the loop.
    if not old_image:
        old_image = new_image
    scaled_image = pygame.transform.scale(new_image, window_size)
    # Only check one frame in 10.
    if count == 10 :
        if check_for_movement(old_image, new_image):
            led_red()
        count = 0
```

```
old_image = new_image
screen.blit(scaled_image, (0, 0))
pygame.display.update()
```

The count variable keeps track of how many times the loop has run. When count gets to 10, the last two images are compared. Sampling only one-tenth of the time also speeds up the program, which would otherwise be too slow. If there was movement, meaning check_for_movement returns True, the LED turns red.

The last part of the main loop checks for the close window event (which stops the program).

```
# Check for events.
for event in pygame.event.get():
    if event.type == pygame.QUIT:
        webcam.stop()
        pygame.quit()
        sys.exit()
    if event.type == pygame.KEYDOWN:
        print(event.key)
        if event.key == 32:  # Space
            led_green()
```

The event checking also catches any key press event (KEYDOWN), and if the spacebar is pressed, the program sets the LED back to green.

USING THE WEBCAM

To get the webcam started, run *monitor.py* by entering the following commands in a terminal window on your Raspberry Pi. A window should open showing a view from the webcam (Figure 5-8).

```
$ cd "/home/pi/zombies/Raspberry Pi/usb_webcam"
$ sudo python monitor.py
```

At this point, the Raspberry Squid LED should be green. To test the movement detection, wave your hand in front of the webcam. The LED should go red and stay red until you press the spacebar on the Raspberry Pi's keyboard.

When everything is working with the webcam connected directly to the Raspberry Pi, you can use the USB extension lead to place the camera further away. Place the camera somewhere overlooking your base's entrance, and then you'll know when the coast is clear to go outside.

There will be a limit on how far you can move the webcam before the signal degrades and you start getting errors, so keep the lead under 30 m.

FIGURE 5-8: THE USB WEBCAM IN OPERATION

PROJECT 8: A WIRELESS ZOMBIE SURVEILLANCE SYSTEM

There may be no Internet after the apocalypse, but that doesn't mean you can't set up your own wireless network and attach a Wi-Fi webcam to it. You can use a low-cost webcam for this project (Figure 5-9). With a wireless webcam, you can put even more distance between you and the zombies you're monitoring, making you safer than ever.

Once you set up the camera and a local network, you can view the camera video from the browser on your Raspberry Pi (Figure 5-10) or even a Wi-Fi-equipped tablet or smartphone. What's more, if you buy the right sort of webcam, you'll be able to use software to change the direction the webcam is pointing.

FIGURE 5-9: A LOW-COST WI-FI WEBCAM

FIGURE 5-10: USING A WI-FI WEBCAM WITH THE RASPBERRY PI

All this comes at a cost, of course: Wi-Fi uses quite a lot of power. The wireless router and Wi-Fi webcam are likely to both use between 5W and 10W of power each. You want to turn them on only when needed.

Note that the Raspberry Pi in Figure 5-8 still has the Raspberry Squid attached, even though this project doesn't use the Squid. Leave Project 7's hardware connected, and you can monitor zombies from both cameras!

WHAT YOU WILL NEED

To setup this Wi-Fi webcam, you'll need the Raspberry Pi setup described in "The Raspberry Pi System" on page 83 and these additional items.

ITEM	NOTES	SOURCE
☐ Wi-Fi webcam	Preferably a unit that can rotate ($50)	Computer store, eBay
☐ Wi-Fi router	Low-end unit ($20) operating from 12V DC supply	Computer store, eBay
☐ 2x Ethernet cable	Any length will do.	
☐ 2x 12V adapter lead	2.1 mm jack plug-to-cigarette lighter adapter	Auto parts store

Wi-Fi webcams are available at a wide range of costs. The device I chose is at the low-cost end and while the image isn't fantastic, it's plenty good enough to spot zombies.

The Wi-Fi router is just a normal household router; most homes with Internet access probably have several, and I'll bet you have a spare lying around, too. These devices serve two purposes: first, to connect your devices to the Internet (not going to happen with zombies all over the place) and, second, to make a local area network (LAN) to which you can attach wired and wireless devices. We'll use the second function of the Wi-Fi router here.

CONSTRUCTION

This project uses ready-made components, so you don't really have any electronics construction to do. You'll just be connecting components (Figure 5-11).

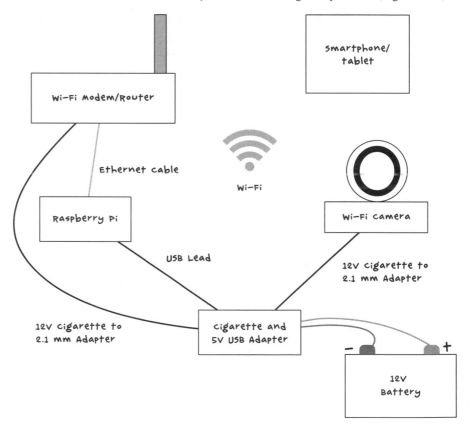

FIGURE 5-11: SCHEMATIC FOR THE WI-FI CAMERA SYSTEM

Connecting a tablet or smartphone to the Wi-Fi network (Figure 5-11) is by no means essential, but it would allow you to monitor the webcam from a mobile device as well as the screen of your Raspberry Pi.

STEP 1: SET UP A LOCAL AREA NETWORK (LAN)

Since this network will not connect to the Internet, you only need a router. That means even if you have a modem-router combination, you don't need to connect it to a phone line or cable connection.

The router allows devices to connect to it in two ways: using an Ethernet cable or using Wi-Fi. We'll connect the Raspberry Pi using an Ethernet cable because a wired connection is more reliable and uses less power than Wi-Fi.

Once you plug the Raspberry Pi into the router, the Pi should automatically join the network using DHCP (Dynamic Host Configuration Protocol), so you shouldn't need to set it up. At this point, though, you may want to set up the Wi-Fi details of the router. This will involve connecting to the configuration page for your router. The IP address for this page is usually 192.168.1.1, but in my case, it was 192.168.1.254. In other words, check your router documentation. When you know the address of your router's admin page, open the Chromium browser and type that URL into the browser's address bar.

The router admin page should have a wireless, WLAN, or Wi-Fi settings page somewhere. Find this page and set the wireless network name (also called the ESSID) and password (Figure 5-12).

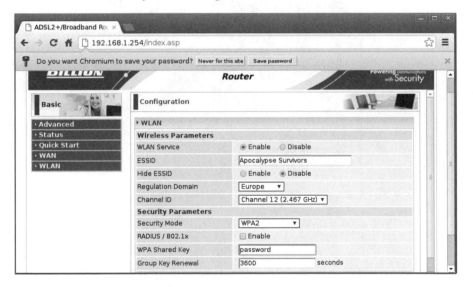

FIGURE 5–12: SETTING UP A WIRELESS NETWORK

Set the network name to something like *Apocalypse Survivors* so tech-savvy survivors can find you easily. Your group of survivors can always benefit from more geeks—especially if it looks like you can run faster than them.

STEP 2: SET UP THE WI-FI CAMERA

The Wi-Fi camera can't connect itself to your wireless network without knowing your password and network name. To give it this information, you'll need to connect to it from a browser, but first it must be connected to the network. This is a bit of a problem. Fortunately, it's a problem that can be resolved by connecting the Wi-Fi camera to the router using an Ethernet cable. Making a wired connection doesn't require a password, and the camera should connect to the network using DHCP just like the Raspberry Pi. After you finish the setup, you can disconnect the Ethernet cable, and the Wi-Fi camera will be free!

Connect the Wi-Fi camera to the router and go back to the same router admin page you used to set up the wireless network. You can use it again to find the IP address of the camera so you can configure it. This time, you're looking for a page called either DHCP table or ARP (Address Resolution Protocol) table. Figure 5-13 shows the ARP table for my router.

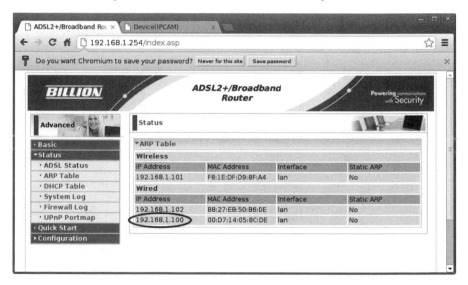

FIGURE 5–13: FINDING THE IP ADDRESS OF THE CAMERA

The connection to the camera is wired, so the IP address of the camera is either 192.168.1.102 or 192.168.1.100. One of those IP addresses belongs to the Raspberry Pi. Find out which is which by entering the **ifconfig** command in LXTerminal. You should see one of the two addresses above in the response to the command, and that's the Raspberry Pi's address.

My Raspberry Pi had an IP address of 192.168.1.102, so by process of elimination, my camera's IP address was 192.168.1.100. Start a new tab on the browser and connect to that IP address, adding :99 after the last number in the address. (I pointed my browser to 192.168.1.100:99.) This extra number specifies the network port to use for the webcam. In most cases, this is 99, but if you're using a different camera, then check your documentation because its port may be different.

Any IP address can have a port number after it like this. Different types of network traffic use different ports. For example, most web traffic uses port 80, which is the default. The webcam happens to use port 99, so this has to be specified in the URL.

The browser should immediately start displaying video from the camera as well as the controls to pan and tilt the camera. Somewhere on the page, you should see a settings link. Click on this and look for Wireless LAN Settings. Click **Wireless LAN Settings**, and you should see an option to scan for Wireless Networks (Figure 5-14).

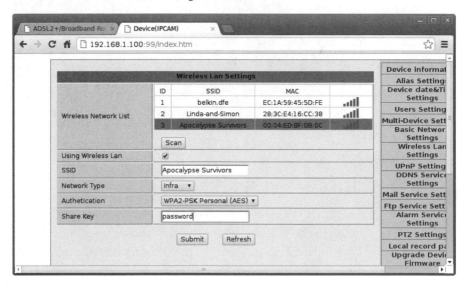

FIGURE 5-14: CONNECTING THE CAMERA TO THE WIRELESS NETWORK

Select the *Apocalypse Survivors* network, enter the password (also called the *share key*), and click **Submit**. The camera should reboot, and then you can unplug the Ethernet because from now on, the camera will use its Wi-Fi connection.

Once the camera has switched over to using Wi-Fi, it will also have a different IP address, so return to the router admin page (Figure 5-13). This time, there should be an entry in the wireless section of the list representing the camera. Try browsing to the camera using that IP address with :99 on the end. Once again, the video should appear in the browser window along with the camera controls (Figure 5-15).

One problem with using DHCP to allocate an address to the webcam is that the router may allocate a different IP address if it is restarted. To avoid this problem, look for the option in your router's DHCP settings that sets the lease time and set this to its maximum. That way, once the IP address is allocated, it shouldn't change until sometime after civilization has been reestablished.

FIGURE 5-15: A VIEW FROM THE WI-FI CAMERA

USING THE WI-FI WEBCAM

Once everything is set up, you can view the image from the webcam by going to the URL for your camera in the browser. The software for most webcams will also allow you to set up multiple cameras and split the screen two or four ways so that you can monitor all the images at once. Then, you can keep tabs on your entrance, supply cache, any zombie traps you've built, and the survivors across the street simultaneously!

You could also access the camera from a mobile browser on a smartphone or tablet computer. There may also be an app for the camera that works better than a browser. This would allow you to work in one area of your compound while keeping an eye on another area using your mobile device. The app provided with the camera I used includes a function to send alerts when movement is detected.

In the next chapter, you'll learn how to control an electric door latch. After completing that project, you'll be able to unlock doors remotely and get inside faster. You'll also be able to detect when the door opens, just in case the undead begin to overrun your base.

6

ADD REMOTE ACCESS AND DETECT OPEN DOORS

Controlling access to your base is key in the postapocalyptic world. Let's say you're being pursued by a herd of brain-hungry zombies. You finally reach your base, and all you have to do to survive another day is get inside to safety. Don't fumble with keys and get eaten before you can unlock the door. Unlock it before you get there! To help you out, this chapter includes projects that allow you to unlock (or lock) doors without touching them. Just don't press any door-opening buttons with without checking your surveillance system first; no matter how politely someone knocks (Figure 6-1), you never know if they're alive or undead.

The first project in this chapter will allow you to open a door by simply pushing a button or even by remote control with a wireless extension. The other project uses a reed switch to detect when a door has been opened and then alerts you using the same Arduino that you used in "Project 4: Battery Monitor" on page 53 and "Project 6: PIR Zombie Detector" on page 72 to monitor the battery voltage and detect zombies with a PIR sensor, respectively.

FIGURE 6-1: POSTAPOCALYPTIC ACCESS CONTROL

PROJECT 9: REMOTE DOOR LOCK

First, let's make reaching the safety of your base a little easier. With an electromechanical door latch, you can press a button to open the door and avoid making jingling key sounds that would attract nearby zombies. This project uses a 12V latch. This door latch will work with the existing door lock, and you can fit one to an existing door by replacing the socket that the lock normally engages with, as shown in Figure 6-2. Note the latch part in the middle that is released by the electromagnet.

FIGURE 6-2: THE ELECTROMECHANICAL DOOR LATCH

The first part of this project builds a simple electrically controlled lock. Press a button to unlock the door (Figure 6-3), and the door will stay unlocked as long as you hold down the button. If you have fellow survivors living with you, this would probably be inside your base, ready for you to let others in. However, if you are on your own, you may want to position it on the outside of your base, right by the door, but high up where it can't be accidentally activated by zombies.

An optional second part of this project lets you use a radio frequency (RF) remote module to unlock the door (see Figure 6-4). A remote-controlled door could save your life, allowing you to run toward your locked door, unlock it just before you get to it, and slam it in the face of that pursuing horde of zombies.

FIGURE 6-3: THE DOOR CONTROL BUTTON

FIGURE 6-4: THE WIRELESS REMOTE CONTROL

WHAT YOU WILL NEED

To make this project, you are going to need the following parts and tools:

ITEM	NOTES	SOURCE
☐ An electric drill and wood bits	You will need larger bit sizes, perhaps up to half inch depending on the width of the door latch.	Hardware store
☐ Hammer	Doubles as a handy weapon	Hardware store
☐ Chisel		Hardware store
☐ Electrical door latch	12V DC	Farnell, Fry's, security store
☐ Fuse	10A fuse and holder	Auto parts store
☐ Push button		Adafruit (1439)
☐ Box for push button		closets, Fry's, garages
☐ Terminal blocks	one three-way block and one two-way block, both 2A	Home Depot, Lowe's, Menards
☐ Double-core wire	Bell wire or speaker cable	Hardware store, scavenge
☐ RF remote switch (optional)	single-channel RF-controlled 12V relay and remote control	eBay

This is one project that requires some woodworking tools. You'll use the drill and set of wood bits, the hammer, and the chisel to make a recess to fit the new door latch, which is generally bigger than normal door latches.

To find a door latch after the apocalypse, you will need to find a specialist security store to scavenge from. Maybe the paper version of the Yellow Pages still has a use! Find your closest security retailer, make your way there carefully, find the latch, and get back to your base. After all, you won't remember what to do with the latch if you become a zombie before you get home.

Almost any double-core wire will work fine, so bell wire or speaker cable is ideal.

CONSTRUCTION

Figure 6-5 shows the schematic for the project. The door latch used in this project remains locked until power is applied to its terminals, and then an electromagnet (electrically powered magnet) releases the latch so the door can open.

This system is great for excluding zombies, but in the event of a fire or other damage to the circuit, this could be very bad: your door would be permanently locked!

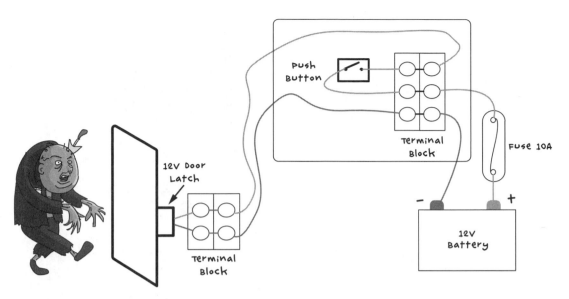

Push Button

Terminal Block

Fuse 10A

12V Door Latch

12V Battery

Terminal Block

FIGURE 6-5: SCHEMATIC FOR THE ELECTRICAL DOOR LATCH

For this reason, any door that you fit this kind of latch to should also retain its original latch; that way you can open it from the inside by twisting the latch. While there won't be anyone around to enforce the fire code, it's not a bad thing to make sure you can get out easily. After all, whatever postapocalypse heating and cooking appliances you cobble together may not be exactly up to code either, making unexpected fires a real possibility.

STEP 1: MAKING A SWITCH BOX

Whether the button is on the inside or outside of your base, you might need to press the door unlock button in a hurry, and letting the button hang about somewhere in a tangle of wires is no good. You need it to be easy to use, so just put it in a box on the wall.

NOTE If you plan to add the wireless control to the door latch, then pick a box that is big enough to contain the remote relay receiver; try placing all of your hardware inside the box to test its size.

Unless you're lucky and manage to scavenge a switch already enclosed in a box, you'll also need to fit the push button into the box. Make sure it has a hole in the lid big enough to mount the switch, as well as holes for the latch and battery lead wires to enter and exit the box. Either find a box with holes large enough or drill the holes out yourself. While you're at it, drill a couple of holes in the bottom of the box to make it easier to fix it to the wall with screws, too. Figure 6-6 shows the switch in a box.

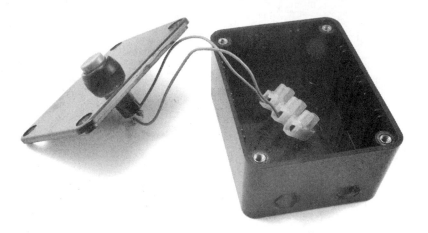

FIGURE 6-6: MAKING A SWITCH BOX. NOTE THE TWO HOLES ON THE SMALL SIDE FACING THE CAMERA, WHICH ARE FOR THE BATTERY AND LATCH WIRES.

Run the switch leads through the hole in your box lid and wire the two terminals of the switch to the terminal block, which will make the overall wiring up of the system easier. The two leads from the switch go to the middle and top positions of the terminal to match the schematic of Figure 6-5.

STEP 2: MAKING THE BATTERY LEAD

To provide power to the project, you need a lead to connect it to your car battery. The lead and fuse shown in Figure 6-7 are just the same as used in "Project 3: LED Lighting" on page 49, so if you need more details on how to make this, have a look at that project.

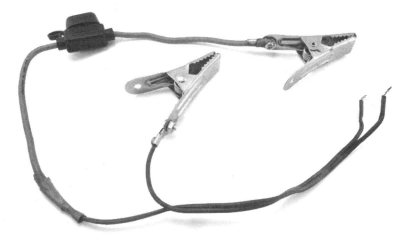

FIGURE 6-7: MAKING A BATTERY LEAD

STEP 3: FITTING THE DOOR LATCH

The electric door latch used in this project is designed to fit into a wooden door frame. If you have a different type of door, search for other 12V door lock mechanisms. Just remember: 12V latches that rely on an electromagnet to keep hold of a metal plate won't keep your base safe. That kind of lock needs to be powered continuously to stay locked, meaning if the battery is empty, your door unlocks and lets all the zombies inside.

To fit the electric door latch, replace the old door latch plate with the electric latch plate. The electric version requires a considerably bigger hole in the door frame to contain the body of the latch, so drill and chisel this hole out; one possible result is shown in Figure 6-8.

FIGURE 6-8: THE LATCH HOLE (A) AND THE FITTED DOOR LATCH (B)

Figure 6-8a shows the latch hole, with a hole drilled at the side to allow the two wires from the lock to be led through to the inside of the door. Figure 6-8b shows the latch fitted back into place. The right edge of the latch releases when power is applied to the latch.

STEP 4: WIRING

Push the ends of the battery lead you made in Step 2 through one of the holes you added to the side of your enclosure in Step 1. Next, wire the positive battery connection to middle position of the three-way terminal block and wire the negative connection to the bottom position.

Unless you're mounting your button right by the door, extend the two wires from the door latch to a reasonable length by joining the short wires of the latch to the longer wire with a two-way terminal block. Then, thread the door latch's long wire through the hole in the back of the switch box and connect it to the top and bottom positions of the screw terminal, as shown in Figure 6-5. When the wiring inside the box is complete, it should look something like Figure 6-9.

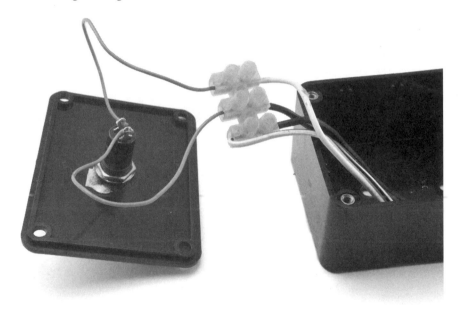

FIGURE 6-9: WIRING UP THE SWITCH BOX

The light-colored wires on the right are for the door latch, and the dark wires are the battery leads. Before you close it all up, just check that pressing the button releases the lock and tidy up the wiring in the box. Finally, affix the door lock's lead to the wall so that it isn't a trip hazard, and you're done!

Of course, your safe haven would be even more accessible if you could unlock the door from a distance, so let's add a remote control.

GOING WIRELESS TO OPEN DOORS AHEAD OF TIME

You could stop after installing a button, but one day, that button won't be fast enough. When you're fresh off a scavenging trip, loaded down with precious supplies and running for your life because a mob of zombies decided to follow you home, you'll wish you could open the door before you reach it. Plan ahead and make the door remote controlled.

To make control of the lock wireless, you can use an RF remote control relay. The relay will be wired in parallel with the push button so if the button is pressed or the remote is activated, the door will unlock.

Figure 6-10 shows the wiring diagram for the project, this time including the wireless remote.

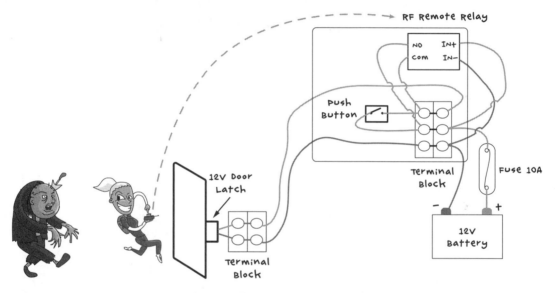

FIGURE 6-10: SCHEMATIC FOR THE ELECTRICAL DOOR LATCH WITH A WIRELESS REMOTE

The push button is connected to the same screw terminals wired to the NO (normally open) and COM (common) connections on the relay. The RF relay module requires a 12V power supply taken from the terminal block's connections to the battery negative and the fuse. Figure 6-11 shows how the relay fits into the same box used for the first part of the project.

FIGURE 6-11: WIRING THE WIRELESS RELAY TO THE ELECTRICAL DOOR LATCH

Wire in the relay according to the diagram in Figure 6-10, and then you'll just need to remember to take your wireless remote with you when you head out to forage or thin out the zombie population. And always bring along a spare remote, or at least a spare battery! As a final backup, you should always keep the real key with you too.

PROJECT 10: DOOR SENSOR

While the first project in this chapter helps you and your loved ones get to safety, the second project alerts you to uninvited guests. Whether a stray zombie or another survivor manages to open the door to your stronghold, with this door sensor, you'll know about perimeter breaches in time to hide.

This project uses a reed switch (if you've never used one, check out "Reed Switches" on page 113 for a detailed description) to detect when a door has been opened, triggering a message on your Arduino. This project uses the same Arduino that monitors your battery and watches for zombies using the PIR detector.

REED SWITCHES

THE SENSOR USED IN THIS PROJECT IS CALLED A <u>reed switch</u>. THIS SWITCH IS MADE FROM A PAIR OF THIN STEEL CONTACTS ENCLOSED WITHIN A SEALED GLASS ENVELOPE. THIS ENVELOPE IS OFTEN FURTHER PROTECTED BY A PLASTIC BOX WITH SCREW HOLES FOR FASTENING IT TO A DOOR OR WINDOW FRAME.

AS SHOWN IN FIGURE 6-12, WITH NO MAGNET PRESENT, THE CONTACTS ARE SLIGHTLY APART, BUT WHEN A MAGNET IS BROUGHT CLOSE, THE TWO CONTACTS ARE PULLED TOGETHER, AND AN ELECTRICAL CONNECTION IS MADE.

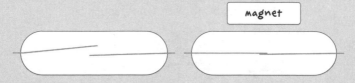

magnet

FIGURE 6-12: A REED SWITCH

BECAUSE REED SWITCHES ARE SEALED, THEY ARE VERY RELIABLE. FOR THIS REASON, THEY'RE OFTEN USED IN SECURITY APPLICATIONS WHERE THE MAGNET IS ATTACHED TO, SAY, THE DOOR ITSELF AND THE REED SWITCH TO THE DOOR FRAME. WHEN THE DOOR IS OPENED, THE MAGNET MOVES OUT OF RANGE OF THE REED RELAY, AND THE CIRCUIT IS BROKEN.

WHAT YOU WILL NEED

To make this project, you're going to need the Arduino and screwshield that you used in "Project 4: Battery Monitor" on page 53, plus a few other parts.

ITEMS	NOTES	SOURCE
☐ Reed switch and magnet pair	After the apocalypse, you can scavenge these from any house that has an intruder alarm.	Adafruit (375), Fry's (1908354), security store
☐ Double-core wire	Speaker cable works well.	Hardware store, scavenge
☐ Terminal block	2-way 2A terminal block	Home Depot, Lowe's, Menards
☐ Arduino	Arduino Uno R3	Adafruit, Fry's (7224833), SparkFun
☐ Arduino screwshield	Screwshield	Adafruit (196)

The reed switch will be further from the Arduino than the short leads that it comes with would allow, so you'll need to extend those leads. Using the double-core wire, either connect the wires together with solder (see "Joining Wires with Solder" on page 231) or connect them to a two-way terminal block.

CONSTRUCTION

Figure 6-13 shows the wiring diagram for connecting the reed switch to the screwshield. You will need the buzzer from "Project 4: Battery Monitor" on page 53, but the resistors are only needed if you also want to monitor the battery voltage.

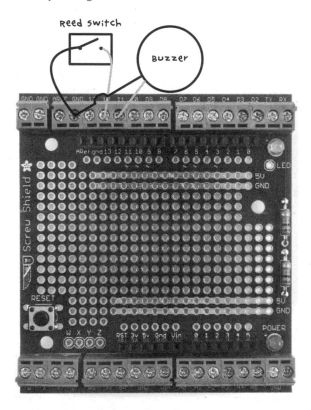

FIGURE 6-13: WIRING DIAGRAM FOR THE DOOR SENSOR

Connect the reed switch to the D12 and GND terminals of the screwshield (it doesn't matter which side goes where), connect the buzzer's positive lead to D11, and connect the buzzer's negative lead to GND. Note that both the negative connection of the buzzer and one connection of the reed switch go to the same GND screw terminal. Figure 6-14 shows the completed project, combined with the resistors used in Project 4.

FIGURE 6-14: THE COMPLETED DOOR SENSOR

The alligator clips at the bottom of Figure 6-14 lead off to the battery, as described in Project 4. With the reed switch hooked up, let's move on to the sketch.

SOFTWARE

All the source code for this book is available online at *http://www.nostarch.com/zombies/*. (See "Installing the Antizombie Sketches" on page 248 for instructions on installing the programs.) If you just want to make this project on its own, without any of the earlier Arduino-based projects, then use the sketch *Project_10_Door_Sensor*. If, on the other hand, you have made one or more of the earlier Arduino projects, then use the sketch *All_Sensors* and change the constants at the top to select the projects that you have made. See the comments section in the *All_Sensors* sketch for instructions on what changes to make.

The code follows the same pattern as Project 4, so for more information on how the program as a whole works, please refer to "Software" on page 57. Here, I will describe just the code specific to this project.

First, a new constant is defined for the Arduino pin that will act as an input for the reed switch.

```
const int doorPin = 12;
```

There is a new line of code in the setup function to initialize that newly defined doorPin (pin 12 on the Arduino) to be an input.

```
pinMode(doorPin, INPUT_PULLUP);
```

The type of input is specified as INPUT_PULLUP so that the input pin will be HIGH by default and only go LOW when the reed switch is closed by being near the magnet. The loop function now also calls a function named checkDoor, which contains the rest of the code for checking for the door being opened.

```
void checkDoor()
{
  if (digitalRead(doorPin))
  {
    warn("DOOR");
  }
}
```

The checkDoor function first reads the door pin. If the result of this read is HIGH, then the magnet is not close enough to the reed switch to hold the switch closed, and the input is in its default state of HIGH. Since the magnet isn't next to the reed switch, the door must be open.

If you only need to know that the door has opened, you don't need a continuous alarm, so checkDoor calls the function warn (passing it "DOOR") rather than alert, which you used for the battery monitor.

```
void warn(char message[])
{
  lcd.setCursor(0, 1);
  lcd.print(message);
  delay(100);
  lcd.setCursor(0, 1);
  lcd.print("            ");
  if (!mute)
  {
    tone(buzzerPin, 1000);
❶   delay(100);
    noTone(buzzerPin);
  }
  delay(100);
}
```

The warn function is like alert: warn takes a message as an argument, prints that message to the LCD, and makes a sound. The difference is that the buzzer tone is cancelled with noTone after just a tenth of a second delay ❶, to give only a short beep when the door is opened.

USING THE DOOR SENSOR

It is always worth testing out a project on your workbench before you install it for real, especially when your life depends on the device working. If this door sensor fails, you could be zombified in your sleep! So first, load your sketch onto the Arduino and line up the reed switch and magnet close together. Then when you move them apart, the buzzer should go off.

Once you're sure everything works as it should, affix the reed switch to the door frame and the magnet to the door. The magnet and reed switch should be opposite each other but not touching. It is best to have the magnet on the door rather than the frame, because the frame doesn't move and will not flex the wires, which would shorten their life. Figure 6-15 shows the reed switch and magnet installed on a door.

FIGURE 6-15: REED SWITCH AND MAGNET ON A DOOR

Note that both the reed switch and magnet are often supplied with adhesive pads on the back to stick them to the door as well as mounting holes, so you can attach them to the wall nonpermanently, as I've done. However, if you are still worried about home decor after the zombie apocalypse, be warned that the adhesive may damage the paint when you remove the reed switch and magnet.

With your new monitor installed, you are ready to take the next step in making your base secure. In Chapter 7, you'll connect smoke and temperature alarms to your hard-working Arduino to protect yourself from more natural disasters that might occur—as if zombies aren't enough!

7

ENVIRONMENTAL MONITORING

Zombies are pretty frightening, but they're not the only threat in a post-apocalyptic world. More mundane risks like fire are especially serious if you can't safely leave your compound (see Figure 7-1). In this chapter, I'll show you how to build a fire alarm and a temperature alarm that alert you to environmental hazards—without alerting the zombies.

FIGURE 7-1: NO SMOKING!

PROJECT 11: QUIET FIRE ALARM

Normally, you want a fire alarm to be as close to you as possible and as noisy as possible. But there's one problem with loud alarms: zombies can hear. The last thing you want when escaping a burning building is to attract unwanted attention from passing zombies!

This project modifies a regular battery-operated smoke detector so that it registers an alarm on the Arduino display and sounds a much quieter buzzer, using the basic setup from "Project 4: Battery Monitor" on page 53. Figure 7-2 shows the smoke detector connected directly to the screwshield.

FIGURE 7-2: TESTING THE FINISHED FIRE ALARM. IN YOUR BASE,
THE DETECTOR WILL BE CONNECTED TO THE ARDUINO BY A LONG LEAD.

WHAT YOU WILL NEED

To make this project, you'll need the Arduino and screwshield that you used in "Project 4: Battery Monitor" on page 53 as well as the following parts:

ITEMS	NOTES	SOURCE
☐ smoke detector	Battery operated	Hardware store, supermarket
☐ cable	Double core and long enough to reach from the smoke detector to the Arduino	Scavenged speaker or bell cable is good for this.
☐ D1	1N4001 diode	Adafruit (755)
☐ R1	1 kΩ resistor	Mouser (293-1k-RC)
☐ LED1	Blue or white LED	Adafruit (301)
☐ C1	100 µF capacitor	Adafruit (753)
☐ solid-core wire	2 inches (5 cm) long	Abandoned electronics, Adafruit (1311)

Be sure to use the LED colors I recommend, as I don't suggest blue or white LEDs just because they look cool. For this project's circuit to work, the LED needs to have a forward voltage of more than about 2V. Red and green LEDs often have a forward voltage of about 1.7V, but blue and white LEDs have a much higher forward voltage of around 3V, which is perfect.

CONSTRUCTION

To adapt the smoke detector to communicate silently with the Arduino, you'll disconnect the detector's buzzer from its circuit board and then change the signal that would go to the buzzer into a signal the Arduino can use. You'll condition the buzzer signal by sending it through the circuit (Figure 7-3).

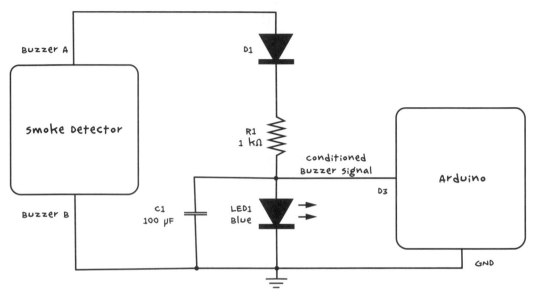

FIGURE 7-3: SCHEMATIC FOR THE FIRE ALARM

A typical smoke alarm is really loud because its buzzer is driven with the highest possible voltage the circuit can wring out of a little 9V battery. This means that for most alarms, the signal on the buzzer looks something like the chart on the left of Figure 7-4.

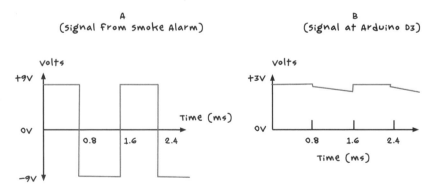

FIGURE 7-4: TAMING THE BUZZER SIGNAL FOR ARDUINO IS MUCH EASIER THAN TAMING A ZOMBIE!

The buzzer is driven by an alternating current (AC) square wave, with a voltage that swings from +9V to –9V at roughly 600 times per second. This causes a piezo element to alternately expand and contract, generating the buzzing sound. But this voltage swing is too wild for the Arduino, which can be damaged by inputs greater than 5V or less than 0V.

The circuit to convert the buzzer signal begins with the diode D1, which completely prevents the negative voltages from reaching the rest of the circuit (diodes only allow current to flow in one direction). The resistor limits the current flowing to the LED, which limits the voltage across the LED to about 3V. The capacitor gets rid of any voltage spikes and smoothes out the signal to something like the chart on the right in Figure 7-4.

STEP 1: DISCONNECT THE BUZZER

First, disassemble the smoke detector. When you remove the lid, you should see a PCB and some wires (Figure 7-5).

FIGURE 7-5: INSIDE THE SMOKE DETECTOR

In this smoke detector, the three leads going from the circuit board to the lid are the buzzer leads. Chop off the leads to the buzzer now, but don't cut too close to the buzzer itself. Resources are scarce during an apocalypse, and you might want to repurpose that buzzer later.

DANGER: RADIATION!

AS YOU START THIS PROJECT, KEEP TWO WARNINGS IN MIND. FIRST, IF YOU TAKE APART YOUR SMOKE DETECTOR BEFORE THE ZOMBIE OUTBREAK, DO NOT USE IT AS A SMOKE DETECTOR AGAIN. SMOKE DETECTORS SAVE THOUSANDS OF LIVES A YEAR, SO DON'T RELY ON ONE YOU'VE MESSED WITH; JUST BUY A NEW ONE.

SECOND, ALTHOUGH REMOVING THE SMOKE DETECTOR'S PLASTIC CASE IS SAFE, IF YOUR SMOKE DETECTOR HAS A ROUND METAL BOX INSIDE (SEE FIGURE 7-5), DO <u>not</u> OPEN THAT BOX, AS IT CONTAINS A RADIATION SOURCE THAT IONIZES AIR IN A SMALL CHAMBER. SMOKE PARTICLES WILL ABSORB THE IONS, AND THE RESULTING REDUCTION IN CURRENT THROUGH THE IONIZED AIR TRIGGERS THE ALARM. THIS TYPE OF SMOKE DETECTOR IS GRADUALLY BEING REPLACED BY DESIGNS THAT DETECT SMOKE OPTICALLY INSTEAD, SO HOPEFULLY, YOURS WON'T HAVE THAT BOX AT ALL.

NOTE *You can use the buzzer from your smoke detector to build "Project 16: Arduino Movement and Sound Distractor" on page 169. If you've ever been close to one of these smoke alarms when they sound, you'll know just how distracting they are!*

Your buzzer may have two leads or three leads. If it has three, follow Step 2 to determine which lead is which. If it has just two, these are the leads that you will connect to, and you can skip Step 2.

STEP 2: IDENTIFY THE LEADS

If your buzzer has three connections, then your smoke alarm uses a type of piezo buzzer called a *self-drive piezo*. The third connection is called the *feedback* connection and is used to make the piezo sound as loud as possible.

For this project, you just want the two drive connections on the smoke detector. Sometimes the wires are color coded; if so, the drive connections will probably be red and black, and the feedback connection might be white (see Figure 7-6) or some other color. But if you have a multimeter, then you can just check which wires are the drive wires and avoid guesswork. Figure 7-6 shows this process in action.

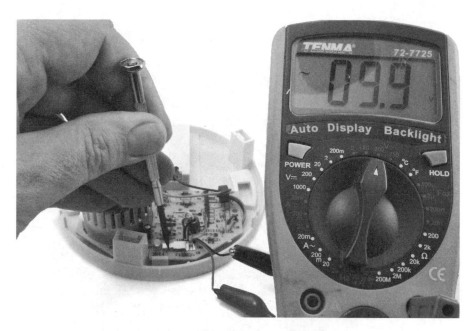

FIGURE 7-6: IDENTIFYING THE SMOKE ALARM BUZZER WIRES

Strip the ends of all three wires and set your multimeter to its 200V AC range if it's available on your meter, or at least the 10V AC range. (Yes, I mean AC, not the usual DC.) Connect the multimeter leads to any two of the three wires and measure the voltage as you hold down the contacts of the smoke alarm's "test" switch. If the meter indicates about 9V, or anything above 4V or 5V, then these are the wires you are looking for; otherwise, try different pairs until you find the correct wires. Note that the project relies on the battery or batteries still being present in the smoke alarm.

STEP 3: SOLDER COMPONENTS TO THE SCREWSHIELD

This circuit has too many components to attach all of them to the screw terminals, so use the prototyping area on the screwshield to solder the components into place. Figure 7-7 shows the wiring diagram for the screwshield; the letters marked will be used later to describe just how to solder this together.

NOTE For the sake of clarity, Figure 7-7 doesn't include components from earlier Arduino-based projects that might be hanging off the screwshield.

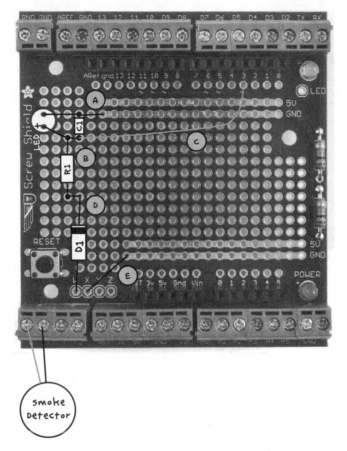

FIGURE 7-7: WIRING DIAGRAM FOR THE SCREWSHIELD

Holding your screwshield so that it looks like Figure 7-7, push the component legs through from the top of the board. Note that the diode (labeled D1) and LED are *polarized*, meaning they only work when oriented a certain way. Point the diode's stripe toward the top of the board. Then place the LED's longer lead (the positive lead) toward the bottom of the board (Figure 7-7).

When you've pushed all the component leads through, flip the board over and solder the leads where they emerge from the hole. (If you are new to soldering, take a look at Appendix B, especially "Soldering a PCB" on page 234.) It may help to bend the leads slightly so that the components don't fall out when the board is upside down. When all the components are soldered, the underside of the board should look like Figure 7-8.

FIGURE 7-8: FIXING THE COMPONENTS IN PLACE

Now that the components are fixed, bend the leads and arrange them to make the connections, using Figure 7-9 and the steps below as a guide. (The connections described below are indicated in Figures 7-7, 7-9, and 7-10 with letters.)

1. Bend the top (negative) lead of the LED over so that it lies next to the top lead of C1 and the GND power line on the screwshield (A). Solder the LED lead where it crosses C1 and then where it meets the GND line. Cut off the excess LED lead and the remainder of the top lead of C2.

2. Bend the other LED lead over to run next to the top lead of R1 and the bottom lead of C1 (B). Solder the bottom LED lead at the junctions where it crosses R1 and C1 and cut off the remainders of both the C1 lead and the R1 lead you just soldered to. If there is any remaining LED lead after connecting to R1 and C1, cut that off too.

3. Cut a length of solid-core wire that is long enough to reach all the way from the end of the positive LED that you soldered in Step 2 as far as D3 on the top Arduino connector (C). Strip the ends of the wire (see "Stripping Wires" on page 227). Flip over to the top side of the board and push one stripped end of the wire into a hole next to where the positive LED lead connects to C1 and solder the wire to that junction. Solder the other end of the wire to the solder pad next to Arduino pin 3. Push the stripped end through the hole from the top and solder on the underside.

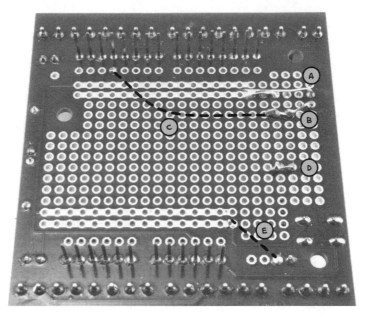

FIGURE 7-9: THE UNDERSIDE OF THE SCREWSHIELD, AFTER SOLDERING.
THE DASHED LINES INDICATE WIRES RUNNING ON THE TOP OF THE SHIELD.

4. Bend the bottom lead of R1 over so that it crosses the top lead of D1 (D). Solder these leads together and cut off the excess wire.

5. Use another short length of solid-core wire (or if they are long enough, one of the leads you trimmed off R1) to connect the solder pad labeled X to the bottom GND power line on the screwshield (E).

When this is done, the underside of the board should look like Figure 7-9. The dotted lines represent the wires on the other side of the board.

Next, flip the board over and add a wire to link pin D3 (just marked 3 on the screwshield) of the Arduino to the junction of the capacitor, diode, and resistor. Solder that wire in place. When this is done, the top of the screwshield should look like Figure 7-10.

Now that the board is complete, reassemble the electronics by fitting the display shield back on top of the screw shield and the screw shield onto the Arduino.

FIGURE 7-10: THE FINISHED SCREWSHIELD

STEP 4: CONNECT THE SMOKE DETECTOR TO THE ARDUINO

Finally, strip the buzzer wires if you haven't done so already, and solder longer leads to them. To make the soldered connections stronger, you could use heatshrink as described in "Using Heatshrink" on page 235. Connect the smoke detector to pins W and X on the screwshield. The wire connecting the smoke detector to the Arduino can be any double-core cable, such as bell wire, but if you plan to use this alarm in your base, just use wires long enough to reach the mounting position. I found that the project worked just fine with 30 feet (10 m) of telephone extension cable.

SOFTWARE

If you want to make this project without any of the other Arduino-based projects in this book, then load the sketch *Project_11_Smoke_Alarm* from this book's source files onto the Arduino now. If on the other hand, you've built one or more of this book's earlier Arduino projects, then use the sketch *All_Sensors* and change the constants at the top to select the projects that you've made. See the comments section in that sketch for instructions on the correct changes to make.

NOTE   You'll find a link to the source code for this book at http://nostarch.com/zombies/. See Appendix C in this book for instructions on loading the programs.

This code builds on the code from Project 4, so for more information on how the program as a whole works, please refer to "Software" on page 57. Here I will just describe the code specific to the fire alarm.

First, we define a new constant for pin D3 on the Arduino:

```
const int smokePin = 3;
```

This pin will act as an input for the signal from the smoke detector. After adding the smokePin constant, we add a new line of code to the setup function to initialize this pin as an input:

```
pinMode(smokePin, INPUT);
```

Next, we add a call to a new function called checkSmoke to the loop function. The checkSmoke function is defined as follows:

```
void checkSmoke()
{
  if (digitalRead(smokePin))
  {
    alarm("FIRE!!");
  }
}
```

The checkSmoke function contains the rest of the code for checking for a signal from the smoke detector and displaying the message and/or turning on the buzzer. To change the display and control the buzzer, call the alarm function, which you first met in "Project 6: PIR Zombie Detector" on page 72:

```
void alarm(char message[])
{
  lcd.setCursor(0, 1);
  lcd.print("          ");
  delay(100);
  lcd.setCursor(0, 1);
  lcd.print(message);
  if (!mute)
  {
    tone(buzzerPin, 1000);
  }
  delay(100);
}
```

Unless you press a button to mute (a holdover from Project 4), this function prints your message ("FIRE!!") to the LCD in lieu of that loud, zombie-attracting buzzer.

USING THE FIRE ALARM

Testing the smoke detector is simple: just hold down the contacts of the test button with a screwdriver (see Figure 7-6). This will cause the buzzer to sound and a message to appear on the LCD screen.

When you know the alarm works, place the sensor somewhere close enough to a potential fire that you'll receive enough advance warning to put out the flames, or at least flee in an orderly manner. Creating a quiet smoke alarm won't be worth much if you exit in a noisy panic and attract all the zombies on the block!

PROJECT 12: TEMPERATURE ALARM

Since your compound is zombie-proofed, you (hopefully) won't have to change lodgings often, and over time, you're sure to acquire some valuable climate-sensitive items. Depending on what you have cached away, you might want to make sure that a generator isn't getting too hot or that your wine cellar isn't too cold. To protect these assets that ensure your survival and are good to trade with other survivors, you need a temperature alarm that can notify you of extremes of heat or cold.

This is the final project that uses your now heavily laden Arduino, and Figure 7-11 shows the LCD screen reporting a high temperature in Celsius.

FIGURE 7-11: A FULLY LADEN ARDUINO, COMPLETE WITH TEMPERATURE SENSOR (CIRCLED), MOVEMENT DETECTOR, SMOKE ALARM, AND BATTERY MONITOR

A three-pin temperature sensor is on the left of Figure 7-11, over the remains of the smoke alarm from Project 11. That sensor will send the Arduino temperature data, which the Arduino will then display as human-readable text.

WHAT YOU WILL NEED

To make this project, you'll need the Arduino and screwshield that you used in "Project 4: Battery Monitor" on page 53 and the following parts:

ITEMS	NOTES	SOURCE
☐ TMP36	Temperature sensor	Adafruit (165)
☐ Three-core wire	To connect the sensor chip to the Arduino screwshield	Scavenged telephone cable or other three-core wire.
☐ Heatshrink	3 lengths of about an inch (25 mm)	Auto parts store

You could use electrical tape instead of heatshrink for this project, but I recommend heatshrink because it's a lot tougher and not prone to unraveling.

CONSTRUCTION

Figure 7-12 shows the wiring diagram for the project. The LCD should be attached from an earlier project, so the only new part you'll add is the TMP36 temperature sensor.

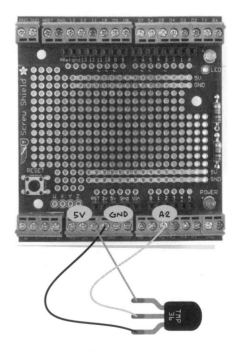

FIGURE 7-12: THE WIRING DIAGRAM FOR THE TEMPERATURE ALARM

TMP36 TEMPERATURE SENSOR

THE TMP36 IS A HANDY LITTLE TEMPERATURE SENSOR CHIP. IT HAS THREE PINS, AND IN THIS PROJECT, THEY'RE CONNECTED TO 5V, GND, AND A2 ON THE ARDUINO. FIGURE 7–13 SHOWS THE PINOUT OF THIS CHIP. THESE CHIPS ARE ONLY ACCURATE TO ABOUT 2 DEGREES CELSIUS. IF YOU WANT GREATER ACCURACY, THEN YOU COULD CONSIDER CHANGING THIS PROJECT'S DESIGN AND SOFTWARE TO USE A DIGITAL TEMPERATURE SENSOR LIKE THE DS18B20.

Vt   out   GND

FIGURE 7–13: THE TMP36 PINOUT

THE POSITIVE SUPPLY VOLTAGE TO PIN Vt ON THE TMP36 CAN BE ANYTHING BETWEEN 2.7V AND 5.5V. ON ITS MIDDLE PIN, THE CHIP PRODUCES AN ANALOG OUTPUT VOLTAGE PROPORTIONAL TO THE TEMPERATURE. THE TEMPERATURE OF THE CHIP (IN DEGREES CELSIUS) CAN BE CALCULATED FROM THE VOLTAGE AT THE OUT PIN BY THIS FORMULA:

$$\text{TEMPERATURE} = 100 \times \text{VOLTS} - 50$$

SO, IF THE VOLTAGE WERE 0.6V, THE TEMPERATURE WOULD BE $100 \times 0.6 - 50 = 10$ DEGREES CELSIUS. IF YOU PREFER YOUR TEMPERATURES IN DEGREES FAHRENHEIT, THEN JUST MAKE ONE FURTHER CALCULATION:

$$°F = °C \times 9/5 + 32$$

THE TMP36 CAN MEASURE TEMPERATURES IN THE RANGE –40 TO +125 DEGREES CELSIUS, BUT THE MEASURED TEMPERATURE IS ACCURATE ONLY TO WITHIN 2 DEGREES CELSIUS OF THE ACTUAL TEMPERATURE.

STEP 1: MAKE A LONGER LEAD FOR THE TMP36

To extend the lead of the TMP36, you could just solder a three-core wire to it. However, to make it a bit more durable, you can use heatshrink tubing on top of the soldered connections. Figure 7-14 shows the process.

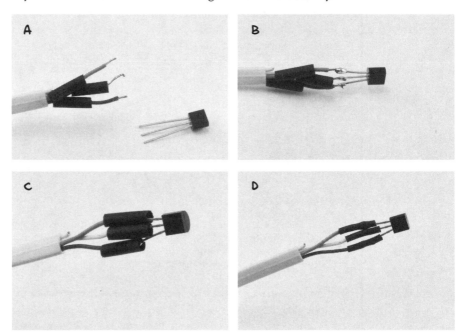

FIGURE 7-14: USING HEATSHRINK ON THE TMP36 LEAD

First, strip the wires of each lead and slip the cut lengths of heatshrink over the individual wires (Figure 7-14a). Next solder the wires to the leads of the TMP36 (Figure 7-14b). Slide the heatshrink up over the solder joint (Figure 7-14c) and finally apply a hair dryer or hot air gun to the heatshrink until it, well, shrinks (Figure 7-14d). If you have wide diameter heatshrink, then you could place this around the whole sensor and individual leads to make this build more durable.

NOTE   For more information on using heatshrink, see the "Using Heatshrink" on page 235.

STEP 2: ATTACH THE TEMPERATURE SENSOR LEAD TO THE SCREWSHIELD

Attach the wires from the temperature sensor to the screwshield (Figure 7-11). You don't have to use the GND connection shown; any of the GND terminals on the screwshield will do.

SOFTWARE

If you want to make this project on its own, without any of the earlier Arduino-based projects, then open the sketch *Project_12_Temperature* from this book's source files and load it on to your Arduino now. If, on the other hand, you built one or more of the earlier Arduino projects, then use the sketch *All_Sensors* and change the constants at the top to select the projects that you have made. See the comments section in this sketch for instructions on this.

NOTE *All the source code for this book is available from http://www .nostarch.com/zombies/. See "Installing the Antizombie Sketches" on page 248 for instructions on installing the programs.*

This code follows the same pattern as Project 4, so for more information on how the program as a whole works, please refer to "Software" on page 57. Here, I'll just describe the code specific to this project.

First, a new constant is defined for the Arduino pin that will act as an analog input for the TMP36:

```
const int tempPin = A2;
```

Two more constants are defined to set the maximum and minimum temperatures allowed before an alarm is triggered. These are floats rather than ints because they represent decimal numbers rather than whole numbers.

```
// Project 12 constants
// these can be in C or F
const float maxTemp = 45.0;
const float minTemp = -10.0;
```

As the comments above the constants state, these temperature values can be in either Celsius or Fahrenheit. The units that the temperature is measured in are decided by a new function you'll define.

The main loop function now includes a call to checkTemp, too. This function is defined as follows:

```
void checkTemp()
{
  float t = readTemp();
  if (t > maxTemp)
  {
    alarm("HOT", t);
  }
  else if (t < minTemp)
  {
    alarm("COLD", t);
```

```
  }
}
```

The checkTemp function first calls readTemp to measure the temperature and then compares that with the maximum and minimum temperatures. If the temperature is too high or too low, then the alarm function is called. Note that this version of the alarm function has an additional parameter that is used to display the temperature on the LCD screen.

The readTemp function is where the raw analog input reading from the TMP36 is converted into a temperature.

```
float readTemp()
{
  int raw = analogRead(tempPin);
  float volts = raw / 205.0;
  float tempC = 100.0 * volts - 50;
  float tempF = tempC * 9.0 / 5.0 + 32.0;
  // One of the following two lines must be uncommented
  // Either return the temperature in C or F
  return tempC;
  // return tempF;
}
```

The raw value returned by analogRead is a number between 0 and 1023, where 0 indicates 0V at the analog input pin and 1023 indicates 5V. This voltage is calculated by dividing the raw value by 205 (205 is roughly 1023/5).

The temperature in degrees Celsius is then calculated using the formula described in "TMP36 Temperature Sensor" on page 133, as the voltage multiplied by 100 with 50 subtracted from the result. The temperature in degrees Fahrenheit is also calculated.

Finally, one of these two values has to be returned. In this version of readTemp, the line to return tempF is commented out, so the temperature in Celsius will be returned. If you want to flip this, then comment out the line return tempC and un-comment return tempF so that the last three lines of the function look like this:

```
  // return tempC;
  return tempF;
}
```

To test the sensor, try changing the value of the maxTemp constant to just above the room's temperature, load the updated sketch onto the Arduino, and then squeeze the temperature sensor between your fingers to warm it up. Watch the LCD, and the readout should change.

USING THE TEMPERATURE ALARM

There's a limit to how much distance you can put between your temperature sensor and your Arduino. You could make the lead you attach to the TMP36 as long as 20 feet (7 m), but the sensor will become less and less accurate as the lead gets longer due to electrical noise on the line and the resistance of the wire.

Leave the sensor near the item you want to stay at a certain temperature and watch the LCD. If that wine cellar just won't stay cool enough, try setting up the sensor in different rooms in your base until you find one with the right climate. If there isn't a good room for the wine, just put the sensor back on your generator, invite the other survivors in your area over for a drink, and have an antizombie strategy meeting.

Now that you have a bunch of sensors to warn you of dangers in your base, in the next chapter, you'll combine the Arduino projects with a Raspberry Pi to make a control center.

8

BUILDING A CONTROL CENTER FOR YOUR BASE

In this chapter, you'll learn how to make an integrated control center using a Raspberry Pi computer interfaced with earlier projects from this book. The control center will allow you to monitor all of your alarm and surveillance devices on one screen so you'll know instantly if a zombie has breached your compound (Figure 8-1). As an extra feature, you'll learn how to add wireless connectivity to your control center.

FIGURE 8-1: A QUIET NIGHT AT THE SECURITY DESK

PROJECT 13: A RASPBERRY PI CONTROL CENTER

In this project, you'll connect the Raspberry Pi system of Chapter 5 with the following Arduino monitoring devices developed earlier in the book:

- "Project 4: Battery Monitor" on page 53
- "Project 6: PIR Zombie Detector" on page 72
- "Project 10: Door Sensor" on page 112
- "Project 11: Quiet Fire Alarm" on page 120
- "Project 12: Temperature Alarm" on page 131

We'll link the two boards with USB cables, which we can later replace in Project 14 with a wireless Bluetooth link. The Arduino will still be able to work without the Raspberry Pi after this wireless modification, but linking it to the Raspberry Pi will allow you to show the status of your sensors and alarms in a window on the Raspberry Pi. Figure 8-2 shows the setup; you can see the sensor status window in the center of the screen.

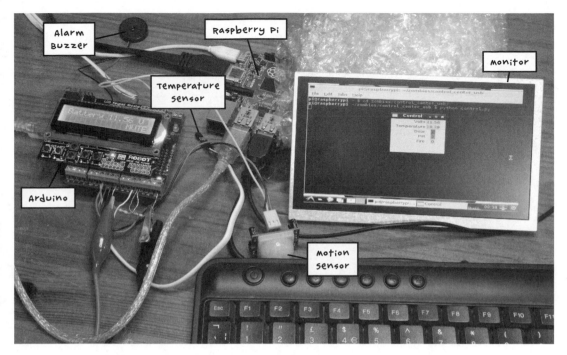

Labels on figure: Alarm Buzzer, Raspberry Pi, Monitor, Temperature Sensor, Arduino, Motion Sensor

FIGURE 8-2: RASPBERRY PI AND ARDUINO WORKING TOGETHER

WHAT YOU WILL NEED

This project brings together the Raspberry Pi system of Chapter 5 and most of the Arduino projects described in the book thus far. As such, all you will need is the following:

☐ One or more of the previous Arduino projects

☐ The Raspberry Pi system from chapter 5

☐ A USB lead/cable (of the same type used to program your Arduino project)

CONSTRUCTION

Assuming that you have been slowly adding projects to your Arduino, the Arduino now has five projects attached to it. If you're really prepared, you probably built these ages ago and have them stashed in your go bag, ready for the apocalypse. Either way, you should at least have the sensors you are interested in using.

If your Arduino projects and Raspberry Pi are already set up, you won't need to do much construction to link them. You connect an Arduino project to the Raspberry Pi by plugging one end of the USB lead into the Pi and the other end into the Arduino. If your Raspberry Pi does not have any free USB ports, then you will need to add a USB hub to provide more ports.

Now that you have linked your Arduino and your Raspberry Pi, you'll need to program them. It's best to program the Arduino from your regular computer before swapping the USB cable over to the Raspberry Pi, as programming the Arduino from the Raspberry Pi's small screen can be frustrating.

Figure 8-3 shows the arrangement of the various system components.

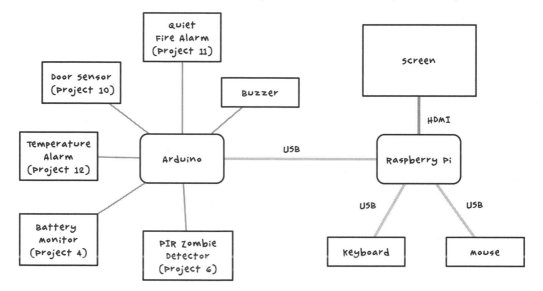

FIGURE 8-3: A SCHEMATIC OF THE CONTROL CENTER

This arrangement plays to the strengths of both the Arduino and Raspberry Pi. The Raspberry Pi cannot directly use many of the sensors that are connected to the Arduino, while the Arduino can. At the same time, the Arduino does not have a screen, while the Raspberry Pi does.

SOFTWARE

There are two parts to the software for this project: a modified version of the *All_Sensors* Arduino sketch and a Python program run on the Raspberry Pi to allow it to communicate with the Arduino.

Before the apocalypse, make sure you've downloaded the source code for this book; go to *http://www.nostarch.com/zombies/* to get started.

ARDUINO SOFTWARE

The Arduino sketch you will use for this project, *Project_13_Control_Center_ USB*, is based on the *All_Sensors* sketch that runs all of the other Arduino projects in this book. *Project_13_Control_Center_USB* just adds code to allow your Arduino to communicate with other devices over a serial connection (in this case, USB).

NOTE For instructions on loading sketches onto your Arduino, see Appendix C.

It's best to test each part of this fairly complex system in isolation on your regular desktop or laptop computer before connecting it to the Raspberry Pi. You can power the Arduino from the USB connection to your laptop while testing, so you don't need to use your postapocalyptic car battery power supply for preapocalyptic testing.

To begin testing, load the *Project_13_Control_Center_USB* sketch onto the Arduino and click the magnifying glass in the Arduino IDE to open the serial monitor (Figure 8-4).

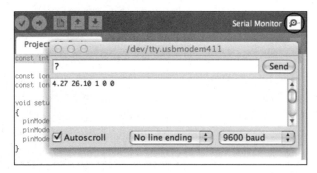

FIGURE 8–4: THE SERIAL MONITOR

Make sure that "9600 baud" is selected in the drop-down list at the bottom right of the serial monitor. This is the baud rate, the speed at which data is sent (measured in bits per second), and it must match the speed set in the sketch.

In the text entry area at the top of the serial monitor, enter the **?** command and click **Send**. The Arduino should display a line of numbers like the 4.27 26.10 1 0 0 shown in Figure 8-4 (your numbers will not match these, exactly). These numbers are the battery voltage, temperature, door status, PIR status, and smoke alarm status, respectively. For the three status values,

0 means everything is okay and 1 indicates an alarm. These are the values that will later be displayed on the control center. By simulating how the Raspberry Pi will fetch the values, you are testing that the Arduino part of the project is working.

If you're currently holding any zombies captive for research, try putting the temperature sensor up against a zombie's skin and enter the ? command again. If you're lacking in test subjects (or feeling less adventurous), just hold the sensor between your fingers. Either way, you should see the temperature part of the message change.

If the responses in the serial monitor indicate that the Arduino side of your control center is working properly, you can unplug the Arduino from the regular computer and attach it to a USB port of the Raspberry Pi.

If the numbers do not appear, then check that the sketch uploaded properly onto the Arduino. If the numbers reported are not what you would expect for one of the projects, then check the wiring for that particular project.

Look at the Arduino code in *Project_13_Control_Center_USB*, and you will see that unlike in *All_Sensors*, the setup function includes the following line at the end:

```
Serial.begin(9600);
```

This line tells the Arduino to open a serial connection, via its USB-serial interface, at a baud rate of 9600. The value passed to begin must match the value you set in the serial monitor's baud rate drop-down list.

This sketch also has a change at the top of the loop function:

```
if (Serial.available() && Serial.read() == '?')
{
  reportStatus();
}
```

These lines check whether any serial communication over USB is waiting to be processed. If so, when you send the ? message, the reportStatus function is called:

```
void reportStatus()
{
  Serial.print(readVoltage());
  Serial.print(" ");
  Serial.print(readTemp());
  Serial.print(" ");
  Serial.print(digitalRead(doorPin));
  Serial.print(" ");
```

```
    Serial.print(digitalRead(pirPin));
    Serial.print(" ");
    Serial.println(digitalRead(smokePin));
}
```

The reportStatus function formats the response from earlier, separating the parts of the message with a space character. The final println command adds a newline character to the end of the response.

RASPBERRY PI SOFTWARE

The program for this project can be found in the *Raspberry Pi/control_center_usb* folder. To download all the Raspberry Pi programs used in this book in one go, you could also use the following commands from a terminal window on the Raspberry Pi:

```
$ cd /home/pi
$ git clone https://github.com/simonmonk/zombies.git
```

These commands should fetch all of the code for the book, including the Arduino code used in other projects.

NOTE   For these commands to work, you will need to have the Pi connected to your network with an Ethernet cable, and your Internet connection needs to be up and running. Therefore, this is definitely something to do when you sense the apocalypse looming. Don't wait until afterward!

To start the control center, you need to run the Python program *control.py*. Enter the following commands in a terminal window on your Raspberry Pi:

```
$ cd "zombies/Raspberry Pi/control_center_usb"
$ python control.py
```

When the program has started up, the window in Figure 8-5 should appear.

The program displays the readings from the Arduino in a human-readable way, and any readings that require your immediate attention will be highlighted in red. When there is no cause for alarm, the readings will be green. In this example, my door is open, which means zombies may be breaking into my compound as I write! While I go check on that, you can open the *control_center_usb.py* file in a text editor and take a look.

FIGURE 8-5: THE CONTROL CENTER

This is the first time we have looked at Python code since chapter 5, so the syntax may look unfamiliar after so much Arduino code. If you get confused about which is which, just keep in mind the main differences: in Python code, there aren't any semicolons at the end of lines, and indentation is used to group code into blocks instead of curly braces as in Arduino code.

I haven't listed the full code for the control center here as it is almost 100 lines, but in the following paragraphs, I'll highlight a few key features. It is useful to know how the code works in case you decide you'd like to modify it. You might, for instance, want to improve the display window so it has an extra column for the units used in the measurements. You could even have it display more explicit warnings about fires, detected zombies, and so on to tell you when you need to get out of your base in a hurry. Visit *http://effbot.org/tkinterbook/tkinter-index.htm* to find out more about making fancy user interfaces in Tkinter.

THRESHOLD VALUES

At the top of the file are three constants that may need to be changed:

```
MIN_VOLTS = 11.0
TEMP_MIN = -10.0
TEMP_MAX = 45.0
```

These constants specify the value limits that determine when the results turn red in the control window. In this example, if the voltage drops under 11V, that row will turn from green to red. The same happens if the temperature drops below –10 or rises above 45 degrees Celsius. The units of temperature used in TEMP_MAX and TEMP_MIN come from the units you used in the Arduino sketch. For details on how to switch between degrees Celsius and Fahrenheit, refer to "Project 12: Temperature Alarm" on page 131.

Set your thresholds to appropriate values for your base, taking into account how much advance warning you want if your battery level gets low or the temperature rises.

STATUS LABELS

The following code shows how the labels and results in the user interface are coded, using battery voltage as an example. The code is contained within a class named App, and the user interface is defined in the __init__ initialize method of this App class.

```
Label(self.frame, text='Volts').grid(row=0, column=0, sticky=E)
self.volts_var = StringVar()
self.volts_label = Label(self.frame, textvariable=self.volts_var)
self.volts_label.grid(row=0, column=1)
```

The first line creates the label Volts and positions it using a grid layout at row 0, column 0. The sticky attribute indicates that the field should "stick" to the "east" wall of the layout cell—in other words, be right justified.

The second line defines a special type of variable (StringVar) used by the Tk graphics library that provides the user interface for the program. This variable is assigned to a member variable called volts_var, which is then referenced in the third line when the label for the voltage value is defined. When the value of the volts_var variable changes, the label field will automatically display the new value of volts_var.

Grid layouts divide the window up rather like table cells and allow you to specify the positions of user interface items without having to provide exact coordinates. The grid is arranged as rows numbered from top to bottom, with the topmost row being 0, and as columns numbered from left to right, with the leftmost column being 0. The last line of code for the volts display positions the label on the grid layout at row 0 and column 1 to put it alongside the label Volts.

The code for the other fields displayed in the window is defined in the same way.

Of course, you may want to use more—or less—descriptive labels, so change them to anything you like. For more information on formatting with the Tk graphics library, see *http://tkinter.unpythonic.net/wiki/*.

COMMUNICATING WITH THE ARDUINO

At the end of the _init_ method you will find these two lines:

```
self.ser = serial.Serial(PORT, BAUD, timeout=1)
time.sleep(2)
```

The first of these lines opens serial communication with the Arduino. The second pauses for two seconds to allow the Arduino time to start up before any messages are sent to it.

KEEPING YOUR CONTROL CENTER UPDATED

If the displayed values don't automatically update, your control center is pretty useless. Updating is accomplished with the read_arduino method.

Here is the first part:

```
def read_arduino(self):
    self.ser.write('?')
    volts, temp, door, pir, fire = self.ser.readline().split()
    self.volts_var.set(volts)
    self.temp_var.set(temp)
    self.door_var.set(door)
    self.pir_var.set(pir)
    self.fire_var.set(fire)
```

The read_arduino method first sends the ? command to the Arduino, which responds with a line of values separated by spaces, as you saw when trying out the Arduino code in the serial monitor. The returned string of values is then split up, using the spaces as a delimiter (this is the default delimiter for the .split() function). StringVars associated with each field in the window are then updated in the display.

After the values are updated, the remainder of the read_arduino method sets the color of the fields to red or green as appropriate.

To ensure that the read_arduino method is called at regular intervals, it is necessary to schedule a call to it from the Tk user interface object:

```
def update():
    app.read_arduino()
    root.after(500, update)

root.after(100, update)
```

This code ensures that after 100 milliseconds (1/10 second), the function update will be called. The function update first calls read_arduino and then schedules itself to run again in 500 milliseconds (half a second), meaning that our control center checks all of our sensors every half second. If you're in danger, whether from zombies or environmental hazards, you'll know quickly!

You can run this program at the same time as you run the USB webcam of "Project 7: Monitor Zombies with a USB Webcam" on page 87 by opening two LXTerminal windows and running one program in each terminal window. That way, you can see instantly what might have triggered your alarms.

USING THE CONTROL CENTER

Now, you have a screen that will give you continuous updates on all of the safeguards of your stronghold. Place your control center somewhere you can easily see it, and if you've included all of the components from Figure 8-3, you'll know instantly if your supplies are in danger, if your power supply is running low, and if zombies have breached your perimeter.

If you find that the user interface values do not update, then go back to "Arduino Software" on page 143 and again test the Arduino using the serial monitor by sending the ? command to look for a status response in the serial monitor.

PROJECT 14: GOING WIRELESS WITH BLUETOOTH

The control center of Project 13 is bogged down in wires right now, and you have to keep the Arduino and the Raspberry Pi together. That also means that you'll probably only know that, say, your base has caught on fire once the flames have reached you—and then it will be too late. You can make your control center much more effective by connecting the Raspberry Pi and the Arduino wirelessly over Bluetooth, as we'll do in this project, so your sensors can detect danger before it reaches you and your monitor.

The Raspberry Pi does not have Bluetooth built in, but it will accept a wide range of Bluetooth USB dongles. We'll add Bluetooth to the Arduino using a Bluetooth serial module, shown sticking out at the right in Figure 8-6.

FIGURE 8-6: ADDING BLUETOOTH TO AN ARDUINO

To make this project, first complete "Project 13: A Raspberry Pi Control Center" on page 140 and make sure that everything else is working properly. Then you'll be ready to add the wireless link.

WHAT YOU WILL NEED

To make this project, you are going to need everything from Project 13 plus the following parts:

ITEMS	NOTES	SOURCE
☐ USB Bluetooth dongle	compatible with Raspberry Pi	computer store, eBay
☐ Bluetooth module	HC-06 Bluetooth serial module	eBay
☐ 270 Ω resistor		Mouser (293-270-RC)
☐ 470 Ω resistor		Mouser (293-470-RC)
☐ connecting wire		
☐ Header pins	4-way	Adafruit (392), eBay
☐ screwshield		Adafruit (196)
☐ multistranded or solid-core hookup wire	For making connections on the prototyping area of the screwshield	Adafruit (1311), scavenge
☐ Female-female jumper wires (x4)	(optional) would replace header pins	Adafruit (266)

The hardware for this project can be built onto the screwshield that you have used while building up the various sensor projects (4, 6, 10, 11, and 12) that use a screwshield. The Bluetooth module I used is a Cambridge Silicon Radio (CSR) device. For a list of Bluetooth dongles compatible with the Raspberry Pi, visit *http://elinux.org/RPi_USB_Bluetooth_adapters/*. If you are worried about soldering the Bluetooth module directly to the header pins, then you may prefer to use four female-to-female jumper wires to link the header pins to the Bluetooth module.

NOTE *You can save yourself some tricky soldering by looking for a module and adapter pair that already has the module soldered into place.*

A lot of the Bluetooth HC-06 modules have six rather than four pins. The pins you will be using are +5V, GND, TXD, and RXD, so you can ignore the other two. These are usually the outside pins, but do check the pinout names as occasionally some designs swap the pin positions around.

CONSTRUCTION

To enable Bluetooth connectivity for your Raspberry Pi, you only need to attach a USB dongle to your system.

The Arduino requires the aforementioned Bluetooth module and a pair of resistors to divide the 5V signal level of the Arduino to the 3V level expected by the Bluetooth module. Mount the module and resistors to the side of the screwshield's prototyping area not already being used by the fire alarm interface from Project 11.

Figure 8-7 shows the wiring layout for the project. To avoid confusion, Figure 8-7 shows the Bluetooth module attached to a screwshield without any other projects built on it.

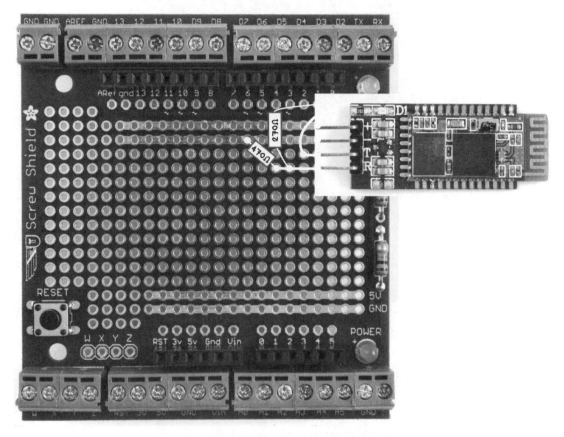

FIGURE 8-7: WIRING LAYOUT FOR ADDING BLUETOOTH TO AN ARDUINO

The Bluetooth module needs to lie flat to keep it out of the way of the LCD shield. For this, you need to solder a row of four 0.1-inch header pins and then solder the Bluetooth module perpendicular to the pins, lying flat over the screwshield. If you prefer, you may also use female-to-female jumper wires to connect the Bluetooth module to the header pins.

STEP 1: SOLDER THE HEADER PINS

Solder the strip of header pins into place. You can see in Figure 8-8 that the +5V and GND pin connections neatly line up with the two power rows at the top of the screwshield.

FIGURE 8-8: THE HEADER PINS SOLDERED IN PLACE

Note that the wire shown leading to pin 3 of the Arduino is part of the fire alarm from Project 11, not this project.

STEP 2: SOLDERING THE RESISTORS AND LINKING WIRE

Solder the resistors and linking wire to the screwshield in the positions shown in Figure 8-9: the 470 Ω resistor goes from GND at Arduino column 7 to the bottom header at column 4; the 270 Ω resistor goes from the bottom pin of the header at row 3 to Arduino pin 1. The connecting wire runs from Arduino pin 0 to the third header pin down.

FIGURE 8-9: SOLDERING THE RESISTORS AND CONNECTING WIRE

When you've soldered the resistors and connecting wire in place, flip the screwshield over to solder the underside of the board.

Figure 8-10 shows a close-up of the underside of the screwshield. To make it easier to identify what is connected to what, the resistors and linking wire are shown as if they were visible through the board.

FIGURE 8-10: CONNECTING THE UNDERSIDE OF THE SCREWSHIELD

First, bend the bottom lead of the 270 Ω resistor over toward the bottom pin header ❶. Solder this to the bottom pin header's pad and snip off the remaining lead. Bend the remaining lead from the bottom end of the 470 Ω resistor to meet the pad one position to its left ❷. Solder the lead to that pad and snip off the excess lead. You have now made a continuous connection from the bottom of the header pins to the bottoms of the 270 Ω resistor and the 470 Ω resistor.

The final connection on the underside ❸ uses the spare wire from soldering the lead from the jumper wire to the header pin to its immediate left.

STEP 3: SOLDERING THE BLUETOOTH MODULE

The final step is to solder the Bluetooth module to the header pins. Solder one pad on the module to one of the header pins, and while keeping the solder molten, position the Bluetooth module so that it is resting against the 1 kΩ resistor that came attached to the screwshield. Then attach the first prong of the module to the first pin. You can see this resistor on the bottom right of Figure 8-9. Once the first prong is soldered, all the other prongs should be lined up and easy to solder. If you prefer, you could use female-to-female jumper wires to link the screwshield to the Bluetooth module. Figure 8-11 shows the Bluetooth module in position.

FIGURE 8-11: THE BLUETOOTH MODULE SOLDERED IN POSITION

SOFTWARE

Since your sensors aren't changing, you'll use the same Arduino software as in "Arduino Software" on page 143. The Bluetooth module replaces the USB interface.

Note that this hardware communicates with the Bluetooth module using the serial port, which on an Arduino Uno is shared with the USB interface. This means that you need to unplug the shield (or just the Bluetooth module if you used jumper wires) before you program the Arduino.

The Raspberry Pi software, however, does need a couple of minor changes, and getting the Raspberry Pi to use Bluetooth does require you to install a whole load of software. Remember: You'll need to install this software before the Internet fails!

Plug the Bluetooth USB adapter into a free USB slot on your Raspberry Pi and then run the following commands in an LXTerminal window:

```
$ sudo apt-get update
$ sudo apt-get install bluetooth
$ sudo apt-get install bluez-utils
$ sudo apt-get install blueman
```

Installing the software will take a considerable amount of time, so you might want to practice your martial arts skills on any willing humans or unwilling zombies available.

When the software is installed and you've worked up a good sweat, reboot the Raspberry Pi with this command:

```
$ sudo reboot
```

Once the Raspberry Pi has rebooted, open a terminal and run the following command to ascertain the ID of the BT interface:

```
$ hciconfig
❶ hci0: Type: BR/EDR  Bus: USB
        BD Address: 00:15:83:0C:BF:EB  ACL MTU: 339:8  SCO MTU: 128:2
        UP RUNNING PSCAN
        RX bytes:419213 acl:19939 sco:0 events:7407 errors:0
        TX bytes:95875 acl:7321 sco:0 commands:57 errors:0
```

The information we want here is the name of the interface, which in this case is hci0 at ❶. When you run this, if the number after hci above is not 0, then make a note of the number; you will need it later.

Every Bluetooth device has a unique ID called a *MAC address*. We need to find the MAC address for our new Arduino Bluetooth module to pair it with the Raspberry Pi. When you power up the Arduino, you should see an LED blinking on the Bluetooth module. The LED is blinking because it has not yet been paired up with the Raspberry Pi; once it has been paired, the LED will go on and stay on. Run the following command to find the ID of the Bluetooth module:

```
$ hcitool scan
```

The output from the hcitool command should look like this:

```
Scanning ...
        00:11:04:08:04:76   linvor
```

The ID is the six-part number. Copy this into the copy-and-paste buffer (Copy and Paste are on the right-click menu). Then enter the following command to link the Raspberry Pi and the Bluetooth module (remember to change the Bluetooth ID to match your Bluetooth module's ID):

```
$ sudo hcitool cc 00:11:04:08:04:76
```

If you have not already done so, follow the instructions in "Raspberry Pi Software" on page 145 for downloading the Raspberry Pi software. You will find the Bluetooth version of *control.py* in the folder *Raspberry Pi/control_center_bt*.

When you have the program, run the following sudo command, again replacing the Bluetooth ID with your own:

```
$ sudo rfcomm connect 0 00:11:04:08:04:76 1 &
[1] 2625
$ Connected /dev/rfcomm0 to 00:11:04:08:04:76 on channel 1
Press CTRL-C for hangup
$
```

You'll need to run this command before you run the program each time your Raspberry Pi reboots. The & on the end of the command runs it in the background so that you can use the terminal window to run the program itself. Hit ENTER to get the $ command prompt back.

If your Bluetooth interface name did not have a 0 after hci when you ran the hciconfig command earlier, change the first 0 after connect to match the number on the end of hci. Remember when I asked you to make a note of this number?

Finally, move to the project directory and run the program:

```
$ cd ~/zombies/control_center_bt/
$ python control.py
```

If you look at the *control.py* files from this project and Project 13, you can see that the only difference is the port. In this version of *control.py*, we set the port to */dev/rfcomm0* rather than */dev/ttyACM0* so that it uses the Bluetooth connection rather than the USB connection.

USING THE BLUETOOTH-ENABLED COMMAND CENTER

The project works in exactly the same way as the USB version in Project 13, with the window displaying the same information, only now it's a little more portable as long as your webcam is wireless. If zombies get into your compound, just grab the Raspberry Pi, monitor, and power source and barricade yourself inside a closet until they lose interest.

In the next chapter, we'll work on ways to distract zombies in a pinch, because the undead are usually much easier to run from than they are to actually kill.

9

ZOMBIE DISTRACTORS

 It doesn't take much to fool a zombie so you can make a quick escape (see Figure 9-1). They are rather lacking in brains, after all. The projects in this chapter are designed to draw zombies' attention away from you using flashes of light, loud sounds, and decoy movements. Imagine you have a herd of zombies lurching around the garage door but you need to get to your last remaining car battery. These distractors will allow you to draw the zombies away from the door or even lure them into a fatal zombie trap, perhaps involving fire and a big hole in the ground.

The first project uses flash units from disposable cameras to produce a disorienting series of flashes to confuse the zombies. The second project uses sound and movement to attract the zombies' attention. Build these projects and affix them to key locations in your base so you can direct zombies away from you.

FIGURE 9-1: SMILE PLEASE!

PROJECT 15: ARDUINO FLASH DISTRACTOR

This flash distractor combines an Arduino and old disposable cameras to produce a timed series of flashes that will confound your brain-hungry foes. Proprietors of old-fashioned photo developer stores are often happy for you to take armfuls of used disposable film cameras off their hands. This is especially true if the proprietors are the animated deceased. They might appear to grumble at you, but I assure you, whatever groaning noises they make are entirely coincidental.

Figure 9-2 shows the completed zombie flash distractor with three salvaged single-use flash cameras, modified to allow the flashes to be triggered by an Arduino. The three cameras are taped together as a block with all the flashes pointing outward.

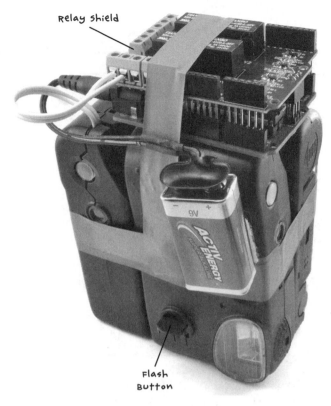

Relay shield

Flash Button

FIGURE 9-2: THE COMPLETED ZOMBIE FLASH DISTRACTOR

The three flash modules are arranged so that each one points at a right angle to every other, giving 270 degrees of coverage. You will need a separate Arduino for this project as you will not want to position this right next to your control center.

WARNING If you have a pacemaker or heart problems, or if flashing lights give you seizures, do __not__ build this project.

WARNING: HIGH VOLTAGES AND BRIGHT FLASHES

FLASHGUNS IN DISPOSABLE CAMERAS OPERATE AT UP TO 400V DC. IF YOU WANT TO AVOID AN UNPLEASANT SHOCK, EXERCISE EXTREME CAUTION WHEN TAKING THE CAMERAS APART AND HANDLING THE FLASH MODULES. MANY PARTS OF THE MODULE WILL BE AT HIGH VOLTAGE AND CAN REMAIN SO FOR HOURS OR EVEN DAYS. BEFORE USING THE MODULES, MAKE SURE TO FOLLOW THE INSTRUCTIONS IN "STEP 3: MAKE THE CAMERA SAFE!" ON PAGE 163 TO SAFELY DISCHARGE THE CAPACITOR.

WHAT YOU WILL NEED

To make this project, you will need the following parts:

ITEMS	NOTES	SOURCE
☐ Arduino Uno	Arduino Uno R3	Adafruit, Fry's (7224833), Sparkfun
☐ Relay shield	4-channel relay shield	eBay, http://www.sainsmart.com/
☐ Disposable cameras	3 used disposable flash cameras	Photo store
☐ 9v battery	PP3 type 9v battery, or larger 9v or 12v battery pack	Hardware store
☐ 9v Arduino battery lead	DC power jack to 9v battery clip adapter	Adafruit (80), eBay
☐ Double-core wire	Three 6-inch (15 cm) lengths of bell wire or other double-core wire	Hardware store, scavenge
☐ Gaffer or electrical tape		Auto parts store
☐ 100 Ω resistor	For discharging the camera's flash capacitor	Mouser (293-100-RC)

There are few uses for spent disposable cameras and a store's only alternative is to pay for someone to take them away, so if you ask store owners nicely, they may give you a stack for free.

Along with a perfectly fine flash module, each camera will generally have an almost unused AA or AAA battery. Try to get a set of cameras that are similar to each other, ideally cameras of the same make. (In the bag of cameras I took away, the most common brand was Fuji, so I based the project on that design. However, the instructions should be sufficiently general to work with any disposable camera.) Also, find cameras that have a switch that turns the flash on for multiple photos, not the sort that make you press the flash button between each shot. For example, look at the camera at the front of Figure 9-2. It has a kind of lever that keeps the flash turned on (bottom center of the figure).

The relay shield was bought on eBay, and when you attach it to the Arduino, it connects a relay to Arduino pins 4, 5, 6, and 7. If you end up with a slightly different relay shield, just check which digital Arduino pins it uses and make the necessary changes in the Arduino sketch (see "Software" on page 166).

The 100 Ω resistor is used to discharge the large, high-voltage capacitor used in the flash module to avoid the risk of electric shocks. It plays no other part in the build.

CONSTRUCTION

Figure 9-3 shows the wiring involved in this project.

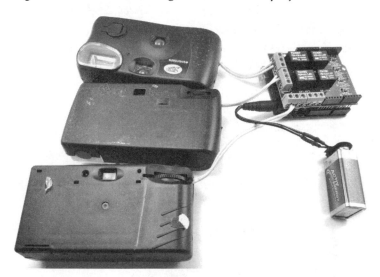

FIGURE 9-3: THE FLASH DISTRACTOR, SPREAD OUT

Each of the disposable cameras has a short length of double-core wire leaving one side of its case. These wires connect to the switch contact inside the camera that is used to trigger the flash. I'll describe how to create this setup for one camera in construction Steps 2 to 5, and you'll need to repeat those steps for all three cameras.

Each pair of leads connects to one pair of relay contacts on the shield so that the Arduino can trigger each flash independently.

Each camera also has its own AA or AAA battery that powers the flash, while the Arduino and relay shield are powered from a 9V battery connected to the DC barrel jack of the Arduino. This makes the project completely portable, so you can place it wherever needed to create a distraction that lets you escape.

STEP 1: SORT THE CAMERAS

First, sort your bag of cameras by type. To make this project simpler to build, try to pick out three identical cameras. The modules I used were all Fujifilm, though the branding on the cardboard covers differed.

WARNING Do not try out the flash of the camera at this stage! It will charge the camera's capacitor, and you'll get shocked later when you lever the camera body apart with your fingers. Seriously, this really hurts!

STEP 2: REMOVE THE TOP CASING FROM A CAMERA

Used cameras may have already been partly disassembled when the photo processor removed the 35mm film canister. The processors do this quickly rather than tidily, so there'll probably be cardboard and bits of plastic hanging off. Figure 9-4 shows the steps involved in taking a camera case apart.

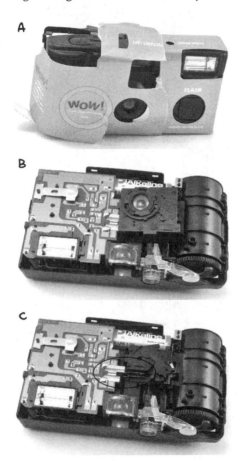

FIGURE 9-4: DISASSEMBLING A CAMERA

You are at risk of shock during this step, so take care not to touch the circuit board or any contacts or wires within the camera.

First, remove the cardboard from the camera body (Figure 9-4a). Next, use a flathead screwdriver with a plastic handle (to provide insulation from shock) to lever apart the plastic catches holding the two halves of the camera body together. Remove the front half of the camera case, exposing the PCB and lens (Figure 9-4b). Now remove the lens assembly. Break it off if you have to; it's not needed anymore. This will expose the two contacts shown circled in Figure 9-4c, which fire the flash when they are touched together.

STEP 3: MAKE THE CAMERA SAFE!

Before you have rendered the camera module safe, treat it the same way you would a small but vicious rodent. Don't handle it directly. If you need to move it around or flip it over, poke it with something like a plastic pen. Otherwise, you might injure yourself, and you need to be in top condition to stay ahead of the zombies.

Identify the flash module's capacitor. The capacitor will be a large metallic cylinder with two leads connecting it to the PCB. The capacitor stores all the energy that is rapidly discharged into the flash to set it off. In Figure 9-5, the entire flash module has been removed from the camera body to make it easier to see the capacitor, but follow the capacitor discharge steps below without removing the whole PCB if you can.

FIGURE 9-5: DISCHARGING THE CAPACITOR

To discharge the flash module's capacitor, bend the legs of your 100 Ω resistor so they are roughly as far apart as the legs of the capacitor. Gently grip the body of the resistor with pliers (with insulated handles) and touch the resistor leads across the capacitor leads. If the capacitor is charged, there will probably be a very small spark. Hold the resistor in place for a second or so to make sure the capacitor actually discharges.

Now, check whether the capacitor is empty by measuring the voltage with your voltmeter set to its maximum DC voltage range. (The voltage range needs to be 500V or more.) It doesn't matter if there are a few volts left in the capacitor, but if you see more than 10V, then discharge it a bit longer with the resistor. Once the voltage is below 10V, it is safe for you to handle the PCB without fear of electrical shock.

STEP 4: ATTACH LEADS TO THE TRIGGER CONTACTS

Solder about 6 inches (15 cm) of double-core wire to the flash contacts, as shown in Figure 9-6. In the camera I used, there was a handy plastic peg that allowed the two contacts to be kept well apart. If this is not the case for your camera, then you may need to wrap the soldered contacts in electrical insulating tape or put heatshrink over the contacts to keep them apart.

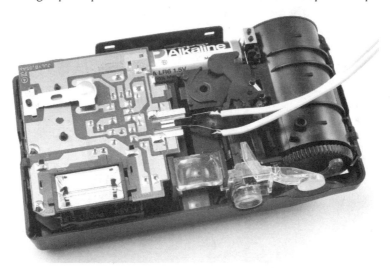

FIGURE 9-6: LEADS SOLDERED TO THE TRIGGER CONTACTS

STEP 5: REASSEMBLE AND TEST THE MODIFIED FLASH MODULE

Fit the front cover of the camera back on, allowing the double-core wire to escape through one side of the camera. If you need more space to snake the wires out, use a pair of diagonal cutters to cut a hole in the plastic cover.

Test this flash before repeating the procedure for the other two cameras. The trigger contacts of these cameras are sometimes at 400V, so for safety, use a screwdriver with an insulated handle.

Turn on the flash switch for the camera. You should see a charging light or LED come on. The camera will probably make a whining noise as the flash charges. This sound is created by the capacitor filling up. When you think the charging is complete (or after, say, 10 seconds), use the screwdriver to connect the two trigger leads, as shown in Figure 9-7.

The camera should flash when you connect the leads with the screwdriver. Hurray! That's one camera ready for action. Before moving on to Step 6, repeat Steps 2 through 5 for the other two cameras.

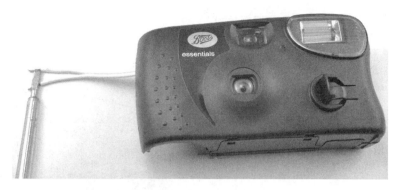

FIGURE 9-7: TESTING THE MODIFIED CAMERA

STEP 6: CONNECT THE CAMERAS TO THE RELAY SHIELD

Fit the relay shield onto your Arduino, making sure that all the pins of the shield engage properly with the sockets on the Arduino.

Figure 9-8 shows how the cameras are wired up to the relay shield.

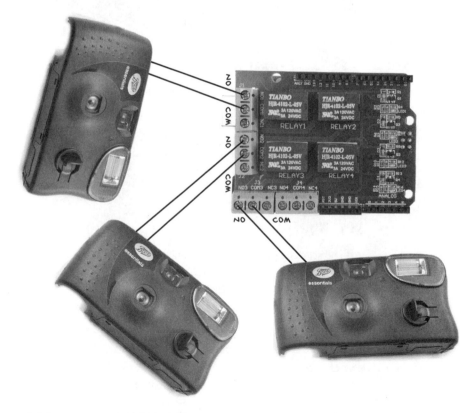

FIGURE 9-8: ATTACHING LEADS FROM THE CAMERAS TO THE RELAY SHIELD'S TRIGGER CONTACTS

Each relay on the relay shield has three screw terminals: NO, COM, and NC. When the relay is not activated, the terminals NC and COM are connected, but when the relay is activated, COM becomes connected to NO. This means that the leads to each camera need to go into the COM and NO connections of each relay. It does not matter which way around the leads go.

When your cameras are connected to their relays, attach the battery clip–to–barrel jack adapter to the Arduino, as shown in Figure 9-9.

FIGURE 9–9: ATTACHING THE BATTERY LEAD TO THE ARDUINO

Before attaching the battery itself, however, you need to upload the software for the project, so you may as well power the Arduino from the USB lead while you program it.

SOFTWARE

All the source code for this book is available at *http://www.nostarch.com/zombies/*. Visit the link provided there and download the code now, if you've not done so already. See Appendix C for instructions on how to install the Arduino sketch.

The Arduino sketch for this project is called *Project_15_Flasher*, and it's in the source file directory of the same name. I'll walk you through this sketch now.

To begin, we define a constant integer array, flashPins:

```
const int flashPins[] = {7, 6, 5};
```

The `flashPins` array defines the Arduino pins used to trigger each of the flash modules. Change these pin numbers if your relay shield uses different pins to control the relays.

Next, we define two more constants, which you can alter to adjust the zombie distractor:

```
const long overallDelay = 20; // seconds
const long delayBetweenFlashes = 1; // seconds
```

The `overallDelay` constant determines how many seconds elapse between each flashing cycle. This value is set to 20 seconds by default. Note, this delay needs to be long enough to enable the capacitor inside the camera to recharge.

The `delayBetweenFlashes` value sets the gap between each of the flashes being triggered in a cycle. This is set to one second by default. Note that both constants are `long` rather than `int`. That's because `int` constants have a maximum value of +/–32,767, which would give a maximum delay of 32.767 seconds; that might not be long enough to keep a zombie distracted while you escape. Fortunately, the `long` data type has a maximum value of over +/–2,000,000. You can run a long way in 2,000 seconds!

Now we add a `setup` function:

```
void setup()
{
  pinMode(flashPins[0], OUTPUT);
  pinMode(flashPins[1], OUTPUT);
  pinMode(flashPins[2], OUTPUT);
}
```

The `setup` function sets all the relay pins to be digital outputs.

With the `pinMode` functions in place, we add a short `loop` function:

```
void loop()
{
  flashCircle();
  delay(overallDelay * 1000);
}
```

This `loop` function calls the `flashCircle` function and waits for `overallDelay` *seconds* before starting the whole process again.

Let's look at the `flashCircle` function definition now:

```
void flashCircle()
{
  for (int i = 0; i < 3; i++)
  {
    digitalWrite(flashPins[i], HIGH);
    delay(200);
```

```
    digitalWrite(flashPins[i], LOW);
    delay(delayBetweenFlashes * 1000);
  }
```

This function loops over the flash pins and gives each a HIGH pulse
for 200 milliseconds, setting off the flash. There is then a pause before the
next flash, set by delayBetweenFlashes. The value of delayBetweenFlashes is
multiplied by 1,000 because in Arduino, the delay function's parameter is in
milliseconds.

USING THE FLASH DISTRACTOR

Before you tape together all the parts of the flash distractor, test it with the
parts laid out as shown in Figure 9-3. Turn on the flash switches of each
camera and attach the 9V battery to the battery clip. The flash units should
flash in turn, before pausing for 20 seconds and then repeating the cycle.

When you know your Arduino can activate the flashes, tape everything
together, or if you prefer, stick the cameras together with a hot glue gun.
Make sure to leave yourself access to the cameras' battery compartments so
you can change the batteries.

The small 9V battery that powers the Arduino will probably last
about four or five hours. If you need to power the distractor for longer,
then Figure 9-10 shows some other options.

FIGURE 9-10: OPTIONS FOR POWERING THE ARDUINO

A 6 × AA battery holder will last around 10 times as long as the PP3 9V battery, but for the ultimate battery duration, you can use one of your car batteries with the cigarette lighter–to–barrel jack adapter shown on the left of Figure 9-10. However, the AA batteries in the cameras won't last more than a few hundred flashes before they need replacing, so if you plan to reuse your distractors—as you might if you're in a zombie-rich neighborhood—keep an extra set of batteries on you.

I suggest you stockpile a few flash distractors and always keep one complete unit in your bag when you venture out for supplies or reconnaissance. Then, if there's a mob of zombies between you and that grocery store you want to scavenge from, just set up the distractor, wait out of sight as it draws the zombies, and when the coast is clear, make a stealthy dash for the doors.

You might want to use the flash distractor in combination with the next project to maximize your ability to distract the undead.

> **NOTE** This project has no on/off switch, so when you are not using it, unplug the 9V battery and turn off the flash switches on the cameras. You could also use an in-line power switch like this: https://www.adafruit.com/products/1125/.

PROJECT 16: ARDUINO MOVEMENT AND SOUND DISTRACTOR

Remember the smoke alarm that we used to make "Project 11: Quiet Fire Alarm" on page 120? In this project, we'll use the piezo buzzer we removed from that smoke alarm, along with a waving flag powered by a servo motor, to make a lot of distracting noise and movement.

Figure 9-11 shows the project in action. Next to the project, I've shown a coiled cigarette lighter adapter, which you can use as an alternative to the AA battery pack if you want to power the setup from a car battery for long-term usage.

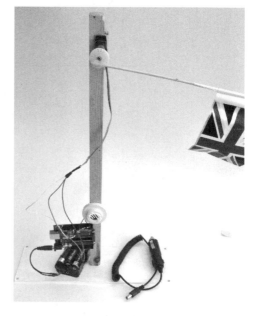

FIGURE 9–11: SOUND AND MOVEMENT IN DISHARMONY

WHAT YOU WILL NEED

To make this project, you will need the following parts:

ITEMS	NOTES	SOURCE
☐ Arduino Uno	Arduino Uno R3	Adafruit, Fry's (7224833), Sparkfun
☐ 6 × AA battery holder	9V battery pack	Adafruit (248)
☐ 9V Arduino battery lead	DC power jack-to-9V battery clip adapter	Adafruit (80), eBay
☐ Jumper wires	3 long male-to-male jumper wires	Adafruit (760)
☐ 100 Ω Resistor	2W or 1/4W	Mouser (594-5083NW100R0J or 293-100-RC)
☐ Pin header	2-way pin header	Adafruit (392), eBay
☐ Servo motor	Small or standard	Adafruit (155 or 196), eBay, hobby store
☐ Buzzer	Discarded from the smoke detector of Project 11 or another high-volume buzzer	security store, smoke alarm
☐ Wooden upright (a post or rod)		Hardware store
☐ Base	Wood or plastic, for attachment of the upright	Hardware store
☐ Wooden food skewer and paper	To make a flag	Household items

The power for this project is supplied through the Arduino barrel jack. The same power options as used in "Project 15: Arduino Flash Distractor" on page 158 also apply to this project.

If you just want to wave a small, lightweight flag like the one shown in Figure 9-11, a small servo motor will work just fine, but for something bigger, use a standard servo. Just be aware that if you use a bigger servo, you may find that your Arduino resets because of voltage drops caused by the load of the bigger motor. In this case, you can power the servo from a separate 6V battery pack as described at *http://communityofrobots.com/tutorial/kawal/how-connect-servo-arduino/*.

Also note that while this project is intended to use the buzzer removed from the smoke detector of "Project 11: Quiet Fire Alarm" on page 120, you can also just use a new buzzer.

CONSTRUCTION

Figure 9-12 shows the wiring diagram for the project.

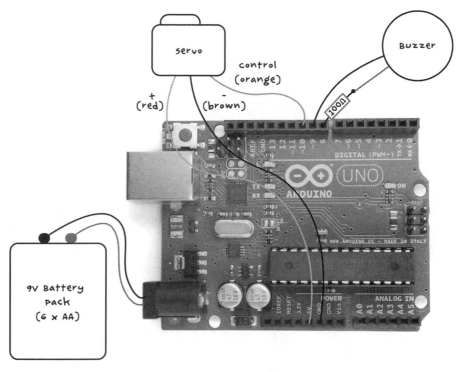

FIGURE 9-12: WIRING DIAGRAM FOR THE DISTRACTOR

The male-to-male jumper wires in the supply list will link the servo motor, which terminates in a three-way socket, to the Arduino. You'll connect the resistor and one buzzer lead to a pair of header pins so that you can plug the buzzer into the Arduino, too.

STEP 1: REMOVE THE PIEZO BUZZER FROM THE SMOKE ALARM COVER

The smoke alarm's buzzer may be integrated into the smoke alarm cover. In that case, don't try to remove the buzzer; you can use it while it's still attached to the cover, or you can just scavenge a different buzzer. If the buzzer looks like it will come away, then remove it as shown in Figure 9-13 to make the project a little more compact.

FIGURE 9-13: REMOVING THE BUZZER FROM THE SMOKE ALARM COVER

STEP 2: SOLDER THE HEADER PINS, BUZZER, AND RESISTOR

Check your buzzer: you only need two buzzer leads, so if it has three, see "Project 11: Quiet Fire Alarm" on page 120 to work out which two of the three leads you need.

Once you have that cleared up, solder the 100 Ω resistor to one buzzer lead—it doesn't matter which one. Solder the other end of the resistor to one of the header pins and the other buzzer lead to the other header pin. You may wish to strengthen these soldered connections using heatshrink (see "Using Heatshrink" on page 231) or electrical tape. These connections are shown in Figure 9-14.

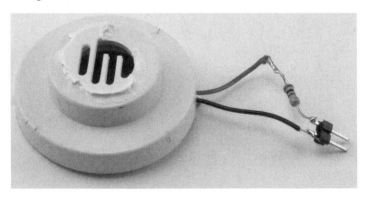

FIGURE 9-14: SOLDERING THE BUZZER, RESISTOR, AND HEADER PINS

STEP 3: TEST THE PIEZO SOUNDER

Before we go further, we'll test the buzzer using the USB connection to power the Arduino. This step will help us find the optimum frequency of the buzzer to make it as loud as possible.

Plug the header pins into the Arduino pins 8 and 9. It does not matter which way around the pins are (Figure 9-15).

FIGURE 9–15: CONNECTING THE BUZZER TO THE ARDUINO

If you haven't downloaded all of this book's programs yet, go to *https://www.nostarch.com/zombies/* and download the *Project_16_Sounder_Test* Arduino sketch. Load this sketch onto your Arduino and then open the serial monitor (Figure 9-16). This is not the final sketch for the project; it's just a test sketch that will let us find the best frequency value to use in the main sketch.

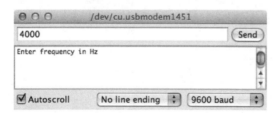

FIGURE 9–16: SETTING THE FREQUENCY USING THE
SERIAL MONITOR

In the input field, enter **4000** and click **Send**—this number is the frequency of the sound your buzzer will emit. You should hear a very loud sound at that frequency for a second. Try entering different frequency values to find the one that gives the highest volume; it will probably be around **4000**.

To reduce the strain on your ears, you can turn the buzzer over or cover the hole where the sound emerges to muffle the volume. When you have found the optimum frequency, make a note of the value.

NOTE   This is definitely something to do when there are no zombies around.

PIEZO BUZZERS

PIEZO BUZZERS (ALSO CALLED SOUNDERS) CONTAIN CRYSTALS THAT CHANGE SHAPE WHEN A CURRENT IS PASSED THROUGH THEM. THE CURRENT CHANGES HUNDREDS OF TIMES PER SECOND, AND AS THE CRYSTALS CHANGE SHAPE, SOUND WAVES ARE PRODUCED. ALTHOUGH YOU CAN DRIVE A PIEZO BUZZER BY CONNECTING ONE LEAD TO GND AND SUPPLYING A SIGNAL TO THE OTHER LEAD, YOU GET A HIGHER VOLUME BY USING TWO ARDUINO OUTPUTS TO COMPLETELY REVERSE THE POLARITY OF THE BUZZER WITH EACH CYCLE. FIGURE 9-17 SHOWS HOW THIS WORKS.

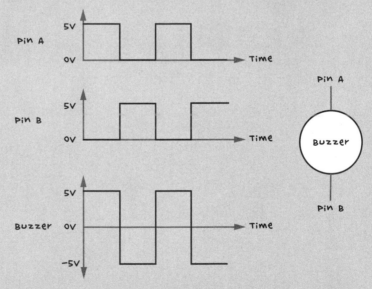

FIGURE 9-17: GENERATING AN ALTERNATING VOLTAGE ON THE PIEZO BUZZER WITH AN ARDUINO. PINS A AND B ARE OUTPUTS ON THE ARDUINO.

WHEN ONE ARDUINO OUTPUT IS HIGH, THE OTHER IS LOW AND VICE VERSA. THIS COMPLETE REVERSAL OF THE POLARITY ACROSS THE PIEZO BUZZER EFFEC-TIVELY ALLOWS A 10V PEAK SWING OF VOLTAGE ACROSS THE BUZZER RATHER THAN THE 5V OBTAINABLE FROM JUST SWITCHING ONE PIN.

YOUR DISTRACTOR WILL NOT BE AS LOUD AS THE ORIGINAL SMOKE DETEC-TOR, WHICH TYPICALLY USES THE SAME TRICK BUT WITH 9V RATHER THAN 5V. HOWEVER, IT SHOULD BE PRETTY LOUD.

STEP 4: MAKE A FLAG

My distractor waves a flag, but yours doesn't have to. Once you have the servo moving, you can attach pretty much anything that will attract the attention of zombies. Try a scrap of rotting meat to get a good zombie-attracting scent going, or if your servo is powerful enough, you might salvage a severed hand for a more realistic human distraction.

Assuming that you just want to wave a flag, the simple arrangement from Figure 9-11 uses a piece of paper folded and glued to a wooden kebab skewer.

STEP 5: ATTACH THE FLAG TO THE SERVO MOTOR

Servo motors generally come with a range of arms and a retaining screw to fix the arm in place on the motor. In this project, I chose the wheel fixture and glued the skewer to it with strong epoxy glue (Figure 9-18).

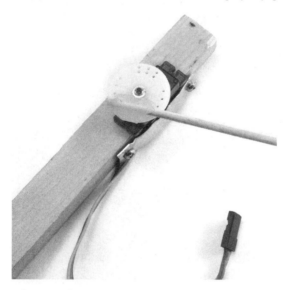

FIGURE 9-18: ATTACHING THE FLAG TO THE SERVO

Don't fit the servo motor's retaining screw just yet, as you will need to adjust the position of the servo arm to accommodate the range of movement (around 160 degrees) once the whole project is up and running.

STEP 6: ATTACH THE SERVO MOTOR TO A BASE

For an upright to attach to the servo motor, I've used a length of wood. To attach the servo, cut a little notch in the wood to fit the servo using a wood saw or small electric hobby cutter. Then use the servo mounting holes and some small screws to fix the servo in place.

As an exercise in ingenuity, I'll leave it up to you to find the best way to attach your servo to your upright. Here, I used a small piece of scrap aluminum to hold the servo in the notch. Epoxy glue would also work.

Now, attach the upright to a base. I drilled a hole in the underside of a flat piece of acrylic and attached the wooden upright with a screw. You may prefer to fix the upright directly to some existing structure, rather than using a freestanding arrangement. (Again, I'll leave the details to your discretion.)

I used the Arduino mounting holes and two more screws to fix the Arduino to the upright as well, but this is entirely optional. Similarly, I stuck the buzzer onto the upright with some glue (Figure 9-19).

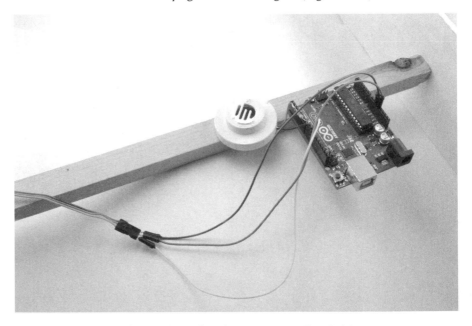

FIGURE 9-19: ATTACHING THE ARDUINO AND BUZZER TO THE UPRIGHT

STEP 7: CONNECTING THE SERVO

Servo motors have leads with three connections that terminate in a single three-hole socket: the black or brown lead is the ground connection, the red lead is the positive power supply, and the third orange or yellow lead is the control signal.

To begin wiring, plug the three male-to-male header leads into the servo's three-hole socket. Run the orange (or yellow) control lead of the servo to pin 10 of the Arduino. Run the black (or brown) ground lead to one of the GND terminals on the Arduino. Finally, connect the red positive supply of the servo to the 5V Arduino pin. Remember: if you are using a large servo, you will probably need an external 6V battery pack, as discussed in "What You Will Need" on page 170.

SOFTWARE

All the source code for this book is available from *http://www.nostarch.com/zombies/*, and the Arduino sketch for this project is called *Project_16_sound_movement*. Download it now and load it onto your Arduino. If you need a refresher on how, follow the directions in Appendix C.

Servos are often used with Arduinos, so there is a built-in library that makes them easy to use together. We import this library at the top of the sketch.

```
#include <Servo.h>
```

Three constants define the behavior of the servo, and tweaking the values of these constants will alter the servo's actions:

```
const int minServoAngle = 10;
const int maxServoAngle = 170;
const int stepPause = 5;
```

Servos have a range of movement of 180 degrees. The constants minServoAngle and maxServoAngle restrict this range between 10 and 170 degrees rather than the full 0 to 180 degrees, because most servos struggle to cover the full 180 degrees.

The constant stepPause sets the delay in milliseconds between each movement of the servo. If you really want to grab a zombie's attention, reduce this number to make the servo move more quickly.

In the next section of code, we define constants for each Arduino pin used.

```
const int sounderPinA = 8;
const int sounderPinB = 9;
const int servoPin = 10;
```

A final constant called f specifies the buzzer frequency:

```
const long f = 3800; // Find f using Project_16_sounder_test
```

Set f to your buzzer's loudest frequency, which you should have noted in "Step 3: Test the Piezo Sounder" on page 173.

Next, to use the servo library, we define a Servo object called arm:

```
Servo arm;
```

With all the constants and global variables defined, we add a setup function to initialize the servo and define the two pins used for the buzzer:

```
void setup()
{
  arm.attach(servoPin);
  pinMode(sounderPinA, OUTPUT);
  pinMode(sounderPinB, OUTPUT);
}
```

The loop function that follows calls two functions to wave the flag and to sound the buzzer:

```
void loop()
{
  wave();
  wave();
  makeNoise();
}
```

The wave function is called twice to waggle the flag back and forth. In case a bit of movement isn't enough to catch a zombie's attention, makeNoise is also called to sound the buzzer. With any luck, the zombies will mistake the noises and movement for something with a brain and will head straight for the distraction!

At the end of the sketch, define the functions that cause the distractions:

```
void wave()
{
  // Wave vigorously from left to right
  for (int angle = minServoAngle; angle < maxServoAngle; angle++)
  {
    arm.write(angle);
    delay(stepPause);
  }
  for (int angle = maxServoAngle; angle > minServoAngle; angle--)
  {
    arm.write(angle);
    delay(stepPause);
  }
}
```

The wave function contains two loops: one loop moves the servo from its minimum to its maximum angle at the preselected speed, and the second loop does the reverse.

Now, let's look at makeNoise:

```
void makeNoise()
{
  for (int i = 0; i < 5; i++)
  {
    beep(500);
    delay(1000);
  }
}
```

This function contains a loop that calls the beep function five times for five beeps of the buzzer. The parameter to beep is the duration of the sound in milliseconds (in this case, 500). Between each beep there is a delay of one second (1000 milliseconds).

NOTE *If using the same values all the time causes local zombies to become immune to your distractor's effects, try tweaking the numbers you pass to beep and delay. You could even randomize the values, using Arduino's random() function.*

The beep function itself generates the AC signal on the two buzzer pins:

```
void beep(long duration)
{
❶  long sounderPeriodMicros = 500000l / f;
❷  long cycles = (duration * 1000) / sounderPeriodMicros / 2;
   for (int i = 0; i < cycles; i++)
   {
     digitalWrite(sounderPinA, HIGH);
     digitalWrite(sounderPinB, LOW);
     delayMicroseconds(sounderPeriodMicros);
     digitalWrite(sounderPinA, LOW);
     digitalWrite(sounderPinB, HIGH);
     delayMicroseconds(sounderPeriodMicros);
   }
}
```

First, we calculate the period of each oscillation ❶ using the frequency f. The resulting value must be further divided by 2 ❷, because what we really want is the duration of delay between swapping the polarity of the pins, and we need two such delays for a complete oscillation.

Using that divided period, the beep function calculates the total number of cycles needed to produce a beep of the correct duration. The for loop that follows uses this information to generate the pulses that are needed.

USING THE SOUND AND MOVEMENT DISTRACTOR

Both of the projects in this chapter need to be kept dry. To weatherproof the sound and movement distractor, you might craft some kind of housing or protective roof shelter. If your build is freestanding, a large plastic bin with a lid could do the trick. I'm sure you can scavenge one from the nearest abandoned discount retail store.

Just cut off one side of the bin so the zombies can see and hear the distractor, affix the project itself to the lid, and snap the bin on top, upside down. Attach this box to a lever and pulley system, and you could even lower it to the ground from the safety of your base, creating a new sport: zombie fishing. Who says you can't have fun during an apocalypse?

Of course, the sound and movement distractor has many practical postapocalyptic uses, too:

- Place it opposite the most vulnerable point in your stronghold to draw away attacking zombie hordes and give you time to reinforce your bunker.
- Plant it by your zombie pit to draw them into the trap.
- Sneak it into your neighbor's yard to thin out the competition for survivor salvage.

In case you want to find other survivors (whether to join forces or avoid them), in the next chapter, we'll look at using wireless technology to communicate.

10

COMMUNICATING WITH OTHER SURVIVORS

In Chapter 1, we discussed the pros and cons of teaming up with other humans when zombies walk the Earth. Associating with other people can certainly be worthwhile: you can protect each other, share knowledge, pool resources, and so on. Of course, they can also take your stuff and put you between themselves and the oncoming zombies. If you decide to take the risk and reach out to your fellow life forms, then build the projects in this chapter.

First, we'll build a beacon to broadcast a voice signal that can be heard on an FM radio, so any survivors scanning the airwaves can hear your message, whether that's "Stay away!" or "Help, I'm trapped on the roof of a shopping mall!" After that, you'll also build a Morse code flasher that will blink out any message you care to translate into dots and dashes.

Of course, if you want to be the one scanning frequency bands, this chapter also explains how to hack a radio receiver to search for a signal. Then, you can lurk silently while you decide whether what's out there is worth broadcasting to (see Figure 10-1).

FIGURE 10-1: ZOMBIES LIKE THE RADIO TOO.

PROJECT 17: A RASPBERRY PI RADIO TRANSMITTER BEACON

The Raspberry Pi is a versatile device that can, given the right software, act as an FM radio transmitter. The only extra hardware you'll need is a length of wire to act as an antenna.

WHAT YOU WILL NEED

This is another Raspberry Pi project, so you will need to have a working Raspberry Pi system complete with keyboard, mouse, and screen as described in Chapter 5. Once the program that transmits the radio signal is up and running, you can turn off the screen to save power if you wish.

RADIO TRANSMITTER LEGALITY

IF YOU'RE READING THIS AFTER THE ZOMBIE APOCALYPSE, THERE WILL BE NO LEGAL PROBLEMS WITH BUILDING A TRANSMITTER BECAUSE THERE WON'T BE ANY GOVERNMENT TO ENFORCE THE REGULATIONS. IF, HOWEVER, YOU ARE BUILDING IN PREPARATION, THEN THE LEGALITY OF THE TRANSMITTER IN THIS PROJECT IS COVERED BY THE SAME LEGISLATION AS FM TRANSMITTERS DESIGNED TO BE CONNECTED TO AN MP3 PLAYER FOR CAR AUDIO.

THESE TRANSMITTERS ARE LEGAL IN THE UNITED STATES IF THE EFFECTIVE RANGE IS 200 FEET (60 M) OR LESS. IF YOU USE A FULL-LENGTH ANTENNA, THIS TRANSMITTER WILL HAVE A LONGER RANGE THAN THAT, SO TO STAY WITHIN THE LAW, USE A SMALL ANTENNA OF ABOUT 3 OR 4 INCHES (7 TO 10 CM).

REGULATION OF THE AIRWAVES IS NECESSARY SO THE FREQUENCIES USED BY EMERGENCY SERVICES STAY CLEAR, BUT THIS TRANSMITTER USES ONLY THE PUBLIC BROADCAST FM WAVE BAND. THE WORST THAT CAN HAPPEN IS ONE OF YOUR NEIGHBORS RECEIVES YOUR BROADCAST INSTEAD OF THEIR FAVORITE RADIO STATION.

To build this radio transmitter, you'll need the following parts:

ITEMS	NOTES	SOURCE
☐ Raspberry Pi	Raspberry Pi 2, Model B or B+	Adafruit (2358), Fry's (8258726)
☐ Jumper wire	Female-to-female jumper wire	Adafruit (826)
☐ Wire for the antenna	About 3 feet (1 m) of wire	

Any wire will do for the transmitter; just check your box of scavenged hookup wire for something that will fit into the end of the female-to-female jumper wire.

You could add the radio transmitter to your existing Raspberry Pi setup. However, for maximum transmission range, you'll want to put the transmitter somewhere high up, so I recommend getting a second Pi.

The length of the jumper wire doesn't matter; it just allows an easy connection between the Raspberry Pi GPIO pin and the antenna wire. The wire to use for the rest of the antenna should be the right size to poke into one end of the female-to-female jumper wire and stay there. You might need to put a kink in the antenna wire so that it stays in place.

CONSTRUCTION

To build your transmitter, all you need to do is plug one end of the jumper wire onto GPIO pin 4 of the Raspberry Pi (Figure 10-2), then plug the antenna wire into the other end of the jumper wire and fix the other end of the antenna to a high spot so that the antenna is pulled up vertically.

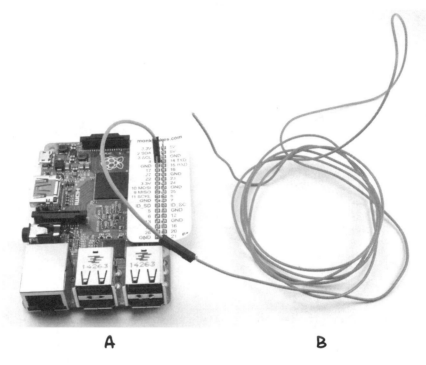

A   B

FIGURE 10-2: ATTACHING THE ANTENNA

You will get the longest transmission range if you place the whole Raspberry Pi up high. If you have a watchtower, this would be ideal.

It does not matter if the antenna wire is not very straight. You may find that some electrical tape wrapped around the junction of the antenna wire and the jumper wire will prevent the antenna from becoming detached. Once you've strengthened the antenna, you've built your radio transmitter beacon!

SOFTWARE

I wish I could claim credit for the wonderful piece of software you'll use in this project, but as it was developed by those clever folk at the Imperial College Robotics Society, I can't. You can find out all about their project at *http://www.icrobotics.co.uk/wiki/index.php/Turning_the_Raspberry_Pi_Into_an _FM_Transmitter.*

The software uses a sound file to oscillate GPIO pin 4 in just the right way to generate an FM carrier wave and signal (see the box on frequency modulation).

To install the software, start an LXTerminal session on your Raspberry Pi and type the following commands:

```
$ mkdir pifm
$ cd pifm
$ wget http://www.icrobotics.co.uk/wiki/images/c/c3/Pifm.tar.gz
$ tar -xzf Pifm.tar.gz
```

These commands create a directory ready to install the software, download the software using the wget utility, and then uncompress the downloaded file into the newly created directory.

USING THE FM TRANSMITTER

To test out the FM transmitter, you need an FM receiver (see "Project 18: Arduino FM Radio Frequency Hopper" on page 188). You also need to find an unused frequency, or at least a frequency with only a faint signal. Of course, this won't be a problem following the apocalypse, but it's more of a challenge with the crowded preapocalypse airwaves. Use your FM receiver to find a quiet part of the spectrum and make a note of the frequency.

The software you installed includes a sound sample of the *Star Wars* theme for testing the transmitter before you record your own, more appropriate message—although the music is not completely inappropriate to accompany humanity's great battle to save itself.

In the LXTerminal, issue the following command to play the tune over your transmitter:

```
$ sudo ./pifm sound.wav 103.0
```

In place of *103.0*, substitute the frequency that your radio receiver is tuned to.

RECORDING A MESSAGE

To record a message, you'll need a laptop and some sound-recording or editing software. I recommend Audacity, which is available free for Windows, OS X, and Linux from *http://audacityteam.org/*.

Fiction and history both tell us that when law and order disintegrate, bad behavior often follows. So think long and hard about what you want to say in your message. Who knows what gun-toting, supply-stealing outlaws are lurking around the corner? You'll probably want to direct new arrivals somewhere you can observe them before lowering your defenses, so bear this in mind when recording your broadcast.

The pifm software requires you to record your message with the sample rate set at 16 bit 44.1kHz and then export the message as a WAV file. In the software, change sound.wav to the name of your new sound file, say *my_message.wav*.

FREQUENCY MODULATION

FREQUENCY MODULATION, OR FM AS IT IS NEARLY ALWAYS CALLED, IS A WAY OF ENCODING A SIGNAL (IN THIS CASE A LOW-FREQUENCY SOUND SIGNAL) ON A MUCH HIGHER CARRIER FREQUENCY. THE SOUND SIGNAL NUDGES THE CARRIER FREQUENCY HIGHER OR LOWER THAN THE CARRIER FREQUENCY, DEPENDING ON THE LEVEL OF YOUR MESSAGE SIGNAL'S WAVEFORM.

FIGURE 10-3 SHOWS TWO CYCLES OF THE MESSAGE SIGNAL (SOLID LINE) SUPERIMPOSED ON THE MUCH HIGHER FREQUENCY CARRIER TO CREATE THE BROADCAST SIGNAL (DOTTED LINE), WHOSE FREQUENCY CHANGES AS YOUR MESSAGE SIGNAL CHANGES.

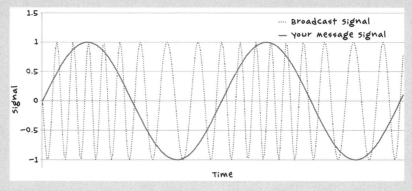

FIGURE 10-3: FREQUENCY MODULATION

WHEN THE SIGNAL IS AT ITS MAXIMUM, THE PEAKS OF THE DOTTED BROADCAST SIGNAL ARE CLOSEST TOGETHER. THAT MEANS THE FREQUENCY IS HIGHER THAN AVERAGE. AT THE BOTTOM OF THE WAVEFORM, WHEN THE SIGNAL HAS ITS MINIMUM VALUE, THE BROADCAST SIGNAL PEAKS ARE FARTHEST APART (THE FREQUENCY IS LOWER THAN AVERAGE).

IN THIS WAY, THE LOW-FREQUENCY SOUND WAVE IS ENCODED ONTO THE HIGH-FREQUENCY CARRIER WAVE. WHEN THIS SIGNAL GETS TO AN FM RADIO RECEIVER, THE CIRCUITRY IN THE RECEIVER EXTRACTS THE ORIGINAL LOW-FREQUENCY AUDIO FROM THE CARRIER SIGNAL.

RUNNING THE TRANSMITTER AUTOMATICALLY

To maximize the chance of other survivors discovering your message, repeat this broadcast around the clock. You can configure the Raspberry Pi to do this for you automatically using a Linux tool called crontab. The crontab utility lets you schedule programs to run at certain times of day.

Enter the following command into the LXTerminal:

```
$ sudo crontab -e
```

This will open a configuration file with the nano editor, as shown in Figure 10-4.

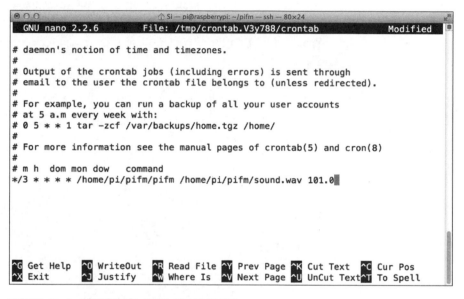

```
GNU nano 2.2.6          File: /tmp/crontab.V3y788/crontab              Modified

# daemon's notion of time and timezones.
#
# Output of the crontab jobs (including errors) is sent through
# email to the user the crontab file belongs to (unless redirected).
#
# For example, you can run a backup of all your user accounts
# at 5 a.m every week with:
# 0 5 * * 1 tar -zcf /var/backups/home.tgz /home/
#
# For more information see the manual pages of crontab(5) and cron(8)
#
# m h  dom mon dow   command
*/3 * * * * /home/pi/pifm/pifm /home/pi/pifm/sound.wav 101.0

^G Get Help  ^O WriteOut  ^R Read File  ^Y Prev Page  ^K Cut Text   ^C Cur Pos
^X Exit      ^J Justify   ^W Where Is   ^V Next Page  ^U UnCut Text ^T To Spell
```

FIGURE 10-4: SCHEDULING YOUR BROADCASTS

Scroll down to the end of the file and add the following line:

```
*/3 * * * * /home/pi/pifm/pifm /home/pi/pifm/sound.wav 101.0
```

The first part of the line (*/3) schedules the transmission to run every 3 minutes, 24 hours a day, 7 days a week. If you use a different sound file or frequency, you need to replace sound.wav with your filename and enter your chosen frequency. If your message is longer than 3 minutes, change */3 to the number of minutes you need it to be.

You only need to do this configuration once; the settings will stick even if the Pi is rebooted.

PROJECT 18: ARDUINO FM RADIO FREQUENCY HOPPER

After the zombie apocalypse strikes, your chances of survival will be increased by group living—that is, assuming no bite victims come inside and turn into zombies. Always be sure that everyone gets checked for zombie-infected wounds before you grant entry!

You'll inevitably need to sleep or go on supply runs, and without someone to watch your back you'll be vulnerable. (Not to mention the slow descent into insanity you'll suffer from lack of human contact—and you thought zombies were crazy.) Therefore, you'll likely benefit from having a few companions around. Other groups of survivors may already be trying to make contact by broadcasting their own radio messages, as we now are. In fact, another group might have bought or salvaged this book and made the FM transmitter of Project 17. To find them, you just need to be able to pick up their transmission.

This project (Figure 10-5) takes a cheap FM receiver and hacks it so that it automatically scans the FM band for the next station. If someone has started transmitting on FM, creating a station instead of the hiss of empty airwaves, you will hear their broadcast. An Arduino simulates the pressing of the tune button on the radio receiver.

FIGURE 10-5: FM RADIO FREQUENCY HOPPER

WHAT YOU WILL NEED

To make this project, you will need the following parts:

ITEMS	NOTES	SOURCE
☐ Arduino	Arduino Uno R3	Adafruit, Fry's (7224833), Sparkfun
☐ FM radio	Simple low-cost FM headphone radio	Dollar store (or equivalently named establishment in your country's currency)
☐ Powered speaker		Electronics store
☐ Audio lead (aux lead)	To connect the radio to the powered speaker	
☐ Red LEDs	2 red LEDs	Adafruit (297)
☐ Barrel jack plug	DC power jack with flying leads, 12V cigarette lighter adapter, or 5V USB adaptor and lead	Adafruit (80), eBay
☐ Right-angle header pins	12-way right-angle header pins	eBay

We are using right-angle pins rather than straight header pins as right-angle pins make it a little easier to solder wires and component leads to this project.

Look for an FM radio that has a Tune button that moves from one station to the next and a Reset button that starts from the beginning of the FM wave band. The radio I used cost less than $2, including in-ear headphones.

The Arduino and speakers both require power. Although I have suggested using the barrel jack, you could just as easily use the USB port to power the Arduino. By now, you should be used to figuring out the most convenient way to power low-voltage devices from a 12V battery.

CONSTRUCTION

This project assumes the radio uses an SC1088 integrated circuit. This extremely low-cost chip is used in most very cheap radios, which seem to use the reference design specified in the datasheet for the chip. (Just search for "SC1088 datasheet" online; you should turn up a PDF in the first few results.) The wiring diagram is shown in Figure 10-6. It shows the Arduino being powered from the DC jack, but it could equally well be powered by the USB port.

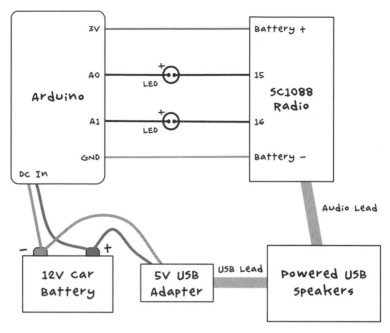

FIGURE 10-6: RADIO SCANNER WIRING DIAGRAM. THE NUMBERS 15 AND 16 ON THE SC1088 RADIO INDICATE PIN NUMBERS OF THE CHIP.

The "tune" and "reset" pins of the SC1088 IC are designed to be connected to momentary pushbuttons that short these pins to the chip's 3V supply rail. You can see this configuration in the datasheet's reference schematic. When pushbuttons are not shorting the input pins to the supply rail, they are pulled down to ground by variable resistances that are set inside the chip. We can emulate the functionality of the pushbutton by connecting these pins to ~3V when we want to simulate a button push, and by leaving the pin *floating* (not being driven high or low) when we want to simulate a button waiting to be pressed. To make the pin float, we can set the Arduino pin that is driving it to an input. When acting as an input, an I/O pin is said to be *high impedance*, meaning that the pin looks like an open circuit to anything that is attached to it.

To convert the 5V of the Arduino output pins to 3V, we place red LEDs between the Arduino pin and the SC1088. These drop the 5V to about 3.3V, the same level as supplied to the chip. The LEDs will also glow very slightly when activated, letting you know when the project is in operation.

STEP 1: DISASSEMBLE THE RADIO

First, take the radio apart. How to do this will depend on how your radio is put together. For mine, I just undid two screws and the whole thing came apart. Figure 10-7a shows the radio in its original state and 10-7b after removal of the case.

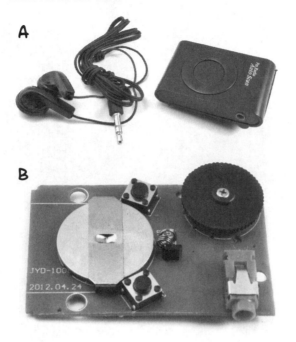

FIGURE 10-7: TAKING THE RADIO APART

Take the button cell battery out because we are going to use the Arduino to supply power to the radio.

STEP 2: IDENTIFY THE CONNECTION POINTS

Now we need to identify the points where we need to attach wires and LED leads. Figure 10-8 shows the underside of the radio's circuit board.

Start by identifying the location of the Scan and Reset switches. The pins for these will form a rectangle. The pins are connected in pairs, so both of the solder points labeled *A* are actually connected, as are the pair of points labeled *B*.

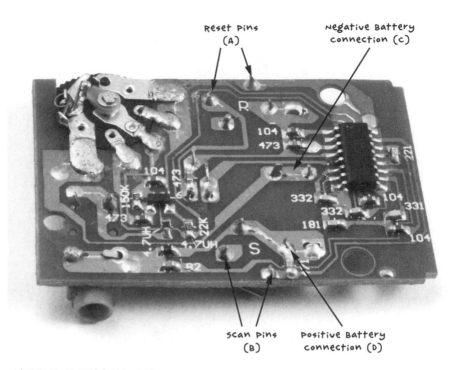

Reset Pins
(A)

Negative Battery
connection (C)

Scan Pins
(B)

Positive Battery
connection (D)

FIGURE 10-8: THE RADIO PCB

The A connections are for the Reset button. If you follow the track on the PCB, you will see that one of the A pins connects to pin 16 of the SC1088 (IC pins are numbered 1 to 16 counterclockwise, with a little dot on the IC package next to pin 1).

Following the track from B, you can see that one pin connects to pin 15 of the SC1088. This is the connection that we will use to scan for the next station.

If you're finding it hard to see where the tracks run, use your multimeter set to continuity mode to identify the pins. Press one probe to the IC pin you want to find a connection for (15 or 16) and then try the different likely connections on the switches with the other probe until the buzzer on the multimeter sounds.

Next, find the two connections needed to power the radio from the Arduino, which correspond to the battery holder connections on the PCB. The 3V batteries the radio takes have a negative central connection (C) and positive connections to the outside frame of the battery holder (D).

STEP 3: ATTACH THE HEADER STRIP

I have suggested a right-angle header strip here, because it's easier to solder the wires to, but regular header pins work almost as well. Break off a length of 12 pins and attach them to the Arduino pins 3.3V through to A5 (Figure 10-9). One pin will sit between the two header sockets, unconnected to anything.

FIGURE 10-9: THE ARDUINO HEADER PINS

STEP 4: LINK THE RADIO TO THE ARDUINO

Figure 10-10 shows the radio connected to the Arduino. Use short wires to connect the 3.3V Arduino pin to the positive battery connection, point D, that you identified earlier. Connect an Arduino GND connection (it doesn't matter which one) to point C, the negative battery connection. Connect the positive (longer) lead of one LED to Arduino pin A0 and the negative lead of that same LED to point B. Do the same with another LED to Arduino pin A1 and point A on the radio PCB.

FIGURE 10-10: THE ARDUINO CONNECTED TO THE RADIO

STEP 5: CONNECT EVERYTHING TOGETHER

Finally, plug the powered speakers into the radio's audio jack. You can test this using the headphones first. The radio uses headphones or an audio lead as an antenna, so you may get better results with a longer lead of a few feet than with a very short lead.

SOFTWARE

All the source code for this book is available from *http://www.nostarch.com/ zombies/*. See Appendix C for instructions on installing the Arduino sketch.

The Arduino sketch for this project is called *Project_18_Scanner,* and I'll walk you through it now.

The sketch starts by defining several constants:

```
const int scanPin = A0;
const int resetPin = A1;const int pulseLength = 1000;
const int period = 5000;
const int numStations = 5;
```

The scanPin and resetPin constants define the two Arduino pins we'll use, and pulseLength defines the length of the simulated button press. The scan buttons needs to be pressed for a full 1,000 milliseconds (1 second) for the radio to scan for the next station rather than simply move the frequency up a step, though this can vary depending on your radio.

The constant period tells the Arduino an amount of time, in milliseconds, to pause so you have time to register whether you are hearing a transmission or just white noise.

Next, we define a single global variable:

```
int count = 0;
```

This variable, called count, is used to keep track of the number of scans to make before resetting to the start of the FM band again.

The setup function initializes both pins as inputs (although as we shall see, this sketch is unusual in that it changes the pin mode of the pins after their first initialization).

```
void setup()
{
  pinMode(scanPin, INPUT);
  pinMode(resetPin, INPUT);
}
```

The loop function is where we actually scan for frequencies:

```
void loop()
{
  delay(period);
  pinMode(scanPin, OUTPUT);
  digitalWrite(scanPin, HIGH);
  delay(pulseLength);
  pinMode(scanPin, INPUT);
  count ++;
  if (count == numStations)
  {
    count = 0;
    pinMode(resetPin, OUTPUT);
    digitalWrite(resetPin, HIGH);
    delay(pulseLength);
    pinMode(resetPin, INPUT);
  }
}
```

First of all, the loop delays by the time specified in period. The function then sends a pulse to the scan pin to begin scanning. When the pulse has finished, the pin is set back as an input.

The `count` variable then increments, and when it has reached the maximum specified in `numStations`, a pulse is sent to the reset pin to start scanning from the beginning of the FM band again. During testing, setting `numStations` to 5 will allow you to check whether the project is working and finding different stations. However, after a zombie apocalypse, the airwaves should be pretty empty, so you may want to reduce this number to just 1, as any signal you happen across is bound to be transmitted by survivors (or perhaps smart zombies). If you discover any automated transmissions you want to ignore, like a distress beacon from your former boss or the murmurings of zombies inexplicably learning the rudiments of human language, change `numStations` to a value of one more than the number of stations you want to ignore.

USING THE RADIO SCANNER

When you first turn everything on, you should hear static. After five seconds or so, the scan LED will glow very dimly, and the radio will scan for its first station. After five more seconds, it will move on to the next station, and so on, until you identify a human friend. Remember: safety in numbers—not hordes.

PROJECT 19: ARDUINO MORSE CODE BEACON

Morse code is a 19th-century invention that allows you to send messages using a series of long or short pulses of light or sound. Each letter of the alphabet is made up of dots and dashes, where a dot is a short pulse and a dash is a long pulse (three times longer than a dot). For example, the letter *z* is represented as this:

z

--..

And the word *zombie* would be this:

zombie

--..　　---　　--　　-...　　..　　.

Morse code uses shorter sequences of dashes and dots for the more commonly used letters, so *e*, as the most common letter used in the English language, is just a single dot. If you are interested, you can search online for the complete Morse code, though the software in this project will translate your message into Morse code for you. Take a look at the code for a table of Morse codes.

This Arduino-based project uses 12V LED lamps, like those you used back in "Project 3: LED Lighting" on page 49, to flash a message to any other survivors in visual range. It's especially effective at night. Figure 10-11 shows the finished project.

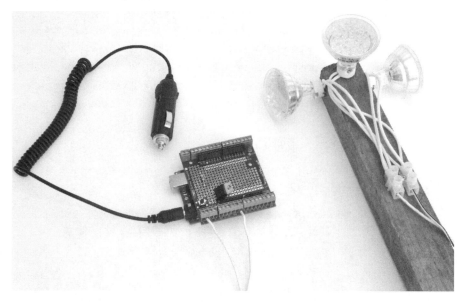

FIGURE 10–11: A MORSE CODE BEACON

WHAT YOU WILL NEED

To make this project, you will need the following parts:

ITEMS	NOTES	SOURCE
☐ Arduino	Arduino Uno R3	Adafruit, Fry's (7224833), Sparkfun
☐ screwshield		Adafruit (196)
☐ 1 kΩ resistor		Mouser (293-1k-RC)
☐ MOSFET	FQP33N10 MOSFET	Adafruit (355)
☐ MR16 LED lamps	12V 3W	Hardware store
☐ MR16 lamp sockets	sockets with trailing leads	Hardware store
☐ Terminal block	2-way terminal block	Home Depot, Lowe's, Menards
☐ 9V Arduino battery lead	DC power jack with flying leads or 12V cigarette lighter adapter	DC power supply
☐ wire	Bell cable (or other cable)	

It is best to use a fresh Arduino and screwshield for this project, both because it will be situated away from your main setup and because your screwshield from previous projects is probably pretty full by now. This project will be powered by its own solar power supply and battery (refer to "Project 1: Solar Recharging" on page 26).

I used three LED lights, but if you want more lamps, just add more in parallel. The transistor used to switch the lights is capable of switching up to 20W of lighting but only with a heatsink, so your combined wattage should be kept below 10W. If you made "Project 3: LED Lighting" on page 49, I would just use the same LEDs.

CONSTRUCTION

The layout for the screwshield and wiring schematic are shown in Figure 10-12.

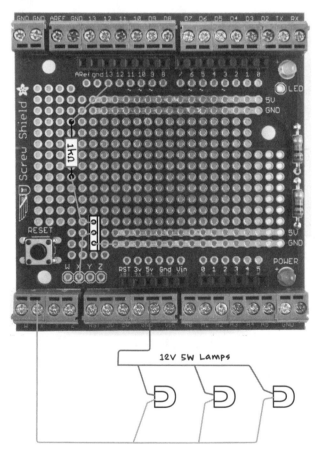

FIGURE 10-12: SCREWSHIELD LAYOUT AND WIRING SCHEMATIC FOR THE MORSE CODE BEACON

STEP 1: ASSEMBLE THE SCREWSHIELD

Assemble the screwshield following the instructions in "Assembling a Screwshield" on page 259.

STEP 2: SOLDER THE COMPONENTS ONTO THE SCREWSHIELD

You only need to solder two components for this project: a resistor and metal oxide semiconductor field effect transistor (MOSFET). MOSFETs are great for switching fairly high-power loads quickly.

Solder the resistor and transistor in place according to the circuit schematic. When soldering the transistor, make sure you place it so that the metal tab faces to the right (Figure 10-12). When the components are soldered into place, the assembly should like Figure 10-13.

FIGURE 10–13: THE TOP OF THE SCREWSHIELD

STEP 3: WIRE THE UNDERSIDE OF THE SCREWSHIELD

Once the components are secured in place, use their excess leads to make the connections on the underside (Figure 10-14). Before soldering the resistor lead that connects to pin 13 on the Arduino, add some insulation to avoid causing short circuits with the 5V and GND tracks it crosses over.

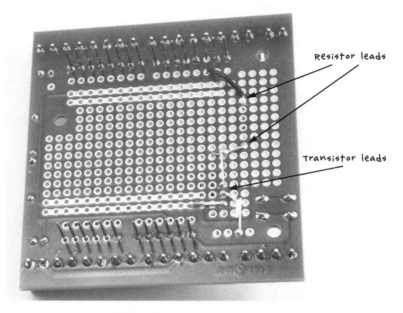

FIGURE 10-14: THE UNDERSIDE OF THE SCREWSHIELD

STEP 4: CONNECT THE LAMPS

If you want to keep this simple, you can just use a single LED lamp. For a wider range of visibility, however, connect a few LED lamps and point them in different directions (Figure 10-15).

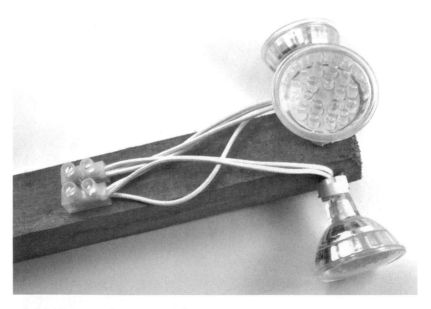

FIGURE 10-15: THE LAMP ASSEMBLY

In Figure 10-15, I've fixed three lamp sockets to a bit of wood and connected all three 12V LED lamps to the terminal block. Lamps of this type usually include a circuit that allows the wires to be connected any way around, but if your modules have a polarity marked on them with a + and –, you need to make sure all the + connections are connected to one terminal of the terminal block and the – connections go to the other. The lamp holders will have holes allowing them to be attached to the wood with screws.

STEP 5: FINAL WIRING

Use some bell cable or other wire to connect the lamp assembly to the X and V_{in} terminals on the screwshield. Stranded wire is best, as it's less liable to break. Make this wire as long as you need it (but above 50 ft, or 15 m, there might be some reduction in brightness): you may want to site the lamp assembly high up outside, to make it easier for people to see your message, while leaving the Arduino in the safety of your bunker. Remember to waterproof the lamp assembly—sealing it in a transparent plastic bag will do the trick.

To connect power to the Arduino, use either a cigarette lighter adapter or a custom lead using alligator clips and a barrel jack plug with flying leads to connect the Arduino to a 12V solar power supply or battery. Note that this project requires 12V for the lamps, so you cannot use a 5V USB lead to power the Arduino.

SOFTWARE

All the source code for this book is available via *http://www.nostarch.com/zombies/*. See Appendix C for instructions on installing the Arduino sketch. The Arduino sketch for this project is called *Project_19_Morse_Beacon*.

The sketch uses the Ardiuno's built-in *EEPROM* library. The Morse code message is stored in EEPROM memory every time a change is made, meaning that the beacon can remember the message even if power to the Arduino is interrupted. The sketch also makes use of a library from the Arduino community called *EEPROMAnything*, which makes saving to and reading from EEPROM easier. The code for EEPROMAnything is included in the download for this project, so there is nothing to download separately.

First, we load both the official Arduino EEPROM library and EEPROMAnything:

```
#include <EEPROM.h>
#include "EEPROMAnything.h"
```

A number of constants are used to control the project:

```
const int ledPin = 13;
const int dotDelay = 100; // milliseconds
const int gapBetweenRepeats = 10; // seconds
const int maxMessageLen = 255;
```

The pin that controls the LEDs is specified in `ledPin`. The constant `dotDelay` defines in milliseconds the duration of a dot flash. Dashes are always three times the duration of a dot.

The constant `gapBetweenRepeats` specifies in seconds the time that will elapse between each repetition of the message, and `maxMessageLen` specifies the maximum length, in letters rather than dots and dashes, of the message. A maximum size is specified because in Arduino code, you have to declare the size of arrays.

Two global variables are used:

```
char message[maxMessageLen];
long lastFlashTime = 0;
```

The `message` variable will contain the text of the message to be flashed, and `lastFlashTime` keeps track of when the message was last flashed, to allow a break between the repeats.

Two global `char` arrays are used to contain the dot and dash sequences for Morse code. The program will only flash characters that it knows how to send, that is letters, digits, or a space character. All other characters in the message are ignored.

```
char* letters[] = {
    ".-", "-...", "-.-.", "-..", ".", "..-.", "--.", "....", "..",    // A-I
    ".---", "-.-", ".-..", "--", "-.", "---", ".--.", "--.-", ".-.",   // J-R
    "...", "-", "..-", "...-", ".--", "-..-", "-.--", "--.."           // S-Z
};

char* numbers[] = {"-----", ".----", "..---", "...--", "....-", ".....",
"-....", "--...", "---..", "----."};
```

The `setup` function sets the `ledPin` as an output and then starts serial communication at `Serial.begin`:

```
void setup()
{
  pinMode(ledPin, OUTPUT);
  Serial.begin(9600);
  Serial.println("Ready");
  EEPROM_readAnything(0, message);
```

```
  if (! isalnum(message[0]))
  {
    strcpy(message, "SOS");
  }
  flashMessage();
}
```

Serial communication is used to set a new message, either using the serial monitor of the Arduino IDE or, as you will see in "Using the Morse Beacon" on page 205, a terminal program running on a Raspberry Pi.

Every time the message is changed, it is saved in EEPROM, so during the setup process, the sketch reads any stored message from EEPROM. If no message has been set, the if statement in setup sets the default message to "SOS." Finally, at flashmessage, the setup function flashes the message for the first time.

The loop function first checks whether a new message has been sent over the serial connection:

```
void loop()
{
  if (Serial.available())        // Is there anything to be read from USB?
  {
    int n = Serial.readBytesUntil('\n', message, maxMessageLen-1);
    message[n] = '\0';
    EEPROM_writeAnything(0, message);
    Serial.println(message);
    flashMessage();
  }
  if (millis() > lastFlashTime + gapBetweenRepeats * 1000L)
  {
    flashMessage();
  }
}
```

Any new message is read into the message character array until the newline character (\n) is read. The null character '\0' is added to the end of the message. This is the Arduino's way of indicating the end of a string of characters. Once the whole message has been read through, it is saved into EEPROM (EEPROM_writeAnything), and then the new message begins flashing immediately.

The remainder of the loop function checks whether enough time has passed before it can repeat the message. This could be done more simply using delay, but we would be unable to interrupt the loop if a new message arrived during the delay.

The flashMessage function is the most complex function in the sketch.

```
void flashMessage()
{
  Serial.print("Sending: ");
  Serial.println(message);
  int i = 0;
  while (message[i] != '\0' && i < maxMessageLen)
  {
    if (Serial.available()) return;   // new message
    char ch = message[i];
    i++;
    if (ch >= 'a' && ch <= 'z')
    {
      flashSequence(letters[ch - 'a']);
    }
    else if (ch >= 'A' && ch <= 'Z')
    {
      flashSequence(letters[ch - 'A']);
    }
    else if (ch >= '0' && ch <= '9')
    {
      flashSequence(numbers[ch - '0']);
    }
    else if (ch == ' ')
    {
     delay(dotDelay * 4);        // gap between words
    }
  }
  lastFlashTime = millis();
}
```

The flashMessage function starts by echoing the message it is about to send to reassure you that it is sending what you want it to. It then loops over every character in the message. Before each character, it uses Serial.available to check for a new message. If a new message has come in, the function stops sending its message in order to receive the new message from your computer or Raspberry Pi; then it begins sending the new message instead.

The flashMessage function determines whether the character is an uppercase letter, a lowercase letter, a number, or the space character and then takes the appropriate action.

If the character is a lowercase letter, the index position of the sequence of dots and dashes held in the letters array is provided as a parameter to the flashSequence function, which then flashes those dots and dashes. The other options are handled in the same way.

Finally, when the whole message has been sent, the lastFlashTime variable is set to the current time so the loop function can work out when it is time to start flashing the message again.

The work of flashing the sequence of dots and dashes for a particular character is handled by the flashSequence function:

```
void flashSequence(char* sequence)
{
    int i = 0;
    while (sequence[i] != NULL)
    {
        flashDotOrDash(sequence[i]);
        i++;
    }
    delay(dotDelay * 3);     // gap between letters
}
```

This loops over each dot or dash, calling flashDotOrDash:

```
void flashDotOrDash(char dotOrDash)
{
  digitalWrite(ledPin, HIGH);
  if (dotOrDash == '.')
  {
    delay(dotDelay);
  }
  else // must be a -
  {
    delay(dotDelay * 3);
  }
  digitalWrite(ledPin, LOW);
  delay(dotDelay); // gap between flashes
}
```

The flashDotOrDash function uses the appropriate delay period to flash a dot or dash.

USING THE MORSE BEACON

Upload the sketch to your Arduino and power up the project. The default message should start to flash. If it doesn't, go back and check over all your wiring. To change the message, attach your Arduino to your computer, open the serial monitor on the Arduino IDE, and type in a new message (Figure 10-16).

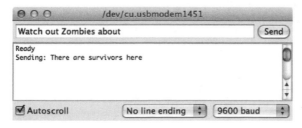

FIGURE 10-16: CHANGING THE MESSAGE USING THE SERIAL MONITOR

Here, the current message, "There are survivors here," should change to "Watch out zombies about" when the Send button is pressed.

If you prefer to use your Raspberry Pi to change the message, install the terminal program screen (your Raspberry Pi will need an Internet connection):

```
$ sudo apt-get install screen
```

Once screen is installed, connect the USB lead between your Raspberry Pi and the Arduino and then enter the following command on your Raspberry Pi:

```
$ screen /dev/ttyACM0 9600
```

At this point, anything you type should be sent to the Arduino, and any messages coming from the Arduino should be displayed. Figure 10-17 shows the message being changed using screen. Note that the message will not appear on the screen as you type it but only after you press ENTER.

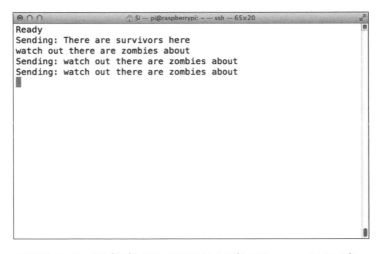

FIGURE 10-17: CHANGING THE MESSAGE USING THE screen COMMAND

Once the message has been changed, the Arduino will remember it, so you can unplug the Arduino to get ready for installation. Unplugging the Arduino will quit the screen command by closing the serial connection to the Raspberry Pi.

Now just attach your project to your desired location, preferably one with 360-degree visibility, and start blinking your message. Figure 10-18 shows the project fixed to my zombie-proof shed.

FIGURE 10-18: INSTALLING THE MORSE BEACON

If you want to conserve power, only use your beacon at night, when it is most likely to be spotted. But beware: popular culture gives us mixed messages on whether zombies are attracted to flashing lights. You may want to reinforce your stronghold before sending out messages, just in case.

In Chapter 11, we will continue with the theme of communication. For the final project of this book, we'll build a pair of haptic communication devices that will allow you and a fellow survivor to communicate silently, without alerting zombies to your presence.

11

HAPTIC COMMUNICATION

If you're out on a supply run, then you'll definitely want this silent communication device, which uses trembling buzzer motors and radio modules to send two-way messages. With this final project, you can communicate without attracting unwanted attention.

PROJECT 20: SILENT HAPTIC COMMUNICATION WITH ARDUINO

The problem with walkie-talkies is that, as the name suggests, they require talking. Zombies have very acute hearing for human speech and will easily home in on any desperate request for backup that you shriek into a walkie-talkie. This is where a silent two-way haptic communication device comes into its own (see Figure 11-1).

FIGURE 11-1: "WHEN THE SIGN SAID 'PRESS FOR ATTENTION,' THIS WASN'T WHAT I THOUGHT IT MEANT!"

Haptic is just a fancy way of saying "relating to touch," and instead of making noise, the devices you'll build in this project will vibrate like a cell phone. You will make a pair of these haptic devices, one of which is shown in Figure 11-2.

Each device has a push-button switch and a small buzzer motor of the sort you find in cell phones. When you press the button on one handset, it causes the buzzer on the other handset to vibrate, and vice versa. The whole thing is powered by a 9V battery.

Then when you are out and about, you can get in touch with your partner using a set of prearranged signals: one short buzz means, "I'm OK"; one long buzz means. "Come quickly, I'm about to be eaten!" In your free time (which has probably increased now that your old school or office is full of zombies), you could even memorize the Morse code you used in "Project 19: Arduino Morse Code Beacon" on page 196 and send more detailed messages.

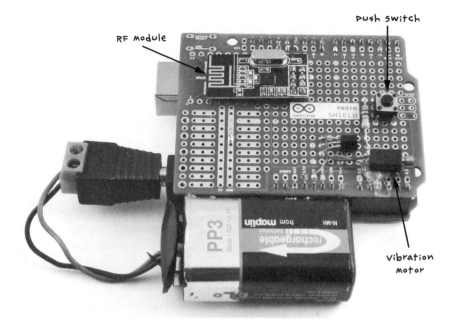

Push switch

RF module

Vibration motor

FIGURE 11-2: A HAPTIC COMMUNICATOR

WHAT YOU WILL NEED

To make this pair of haptic communicators, you'll need the following parts:

ITEMS	NOTES	SOURCE
☐ Arduinos	2 x Arduino Uno R3	Adafruit, Fry's (7224833), SparkFun
☐ Protoshields	2 x Arduino Protoshield PCB	eBay (Arduino code: A000077)
☐ Header pins	Header pins 64 way in total (for 2 handsets)	Adafruit (392), eBay
☐ 9V battery leads	2 x Arduino 9v battery leads	Adafruit (80), eBay
☐ 9V batteries	2 x PP3 batteries	Hardware store
☐ R1	2 x 1 kΩ resistor	Mouser (293-1k-RC)
☐ Transistors	2 x 2N3904 NPN bipolar transistor	Adafruit (756)
☐ Vibration motors	2 x 5V or 3V vibration motor	eBay
☐ Tactile switch	2 x tactile push switch	Adafruit (504)
☐ RF modules	2 x NRF24 RF modules	eBay
☐ Assorted hookup wire	Stranded wire	
☐ Wire	Insulated solid-core wire for making PCB connections	

You might also want to enclose your communicators in plastic boxes to protect them from the elements. If you choose to do so, then you will need to find something big enough to contain the Arduino, protoshield, and battery. It will also need a hole so that you can press the button and add an on/off switch.

Electronically, this is probably the most complicated project so far. You might struggle to find all the parts after a zombie apocalypse, as some, like the vibration motors and the RF modules, are best bought off eBay or Amazon. So make this project now, before the postal service un-dies. Vibration motors can also be scavenged from cellphones.

CONSTRUCTION

These instructions will tell you how to make one haptic module, and Figure 11-3 shows the schematic for one communicator. Of course, to communicate with someone else, you will need to make two devices.

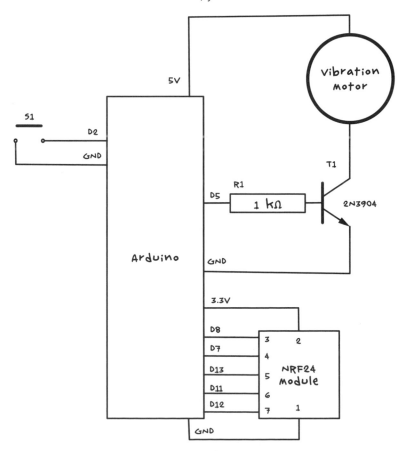

FIGURE 11-3: THE SCHEMATIC FOR ONE HAPTIC COMMUNICATOR

Pin 2 of the Arduino will be set up as a digital input with internal pullup resistor enabled, connected to the push button S1. When the button is pressed, the Arduino will control the NRF24 radio module to send a message to the other handset, activating its vibration motor.

The vibration motor is controlled from pin D5 of the Arduino. We use a transistor (T1) because the motor uses more current than the Arduino output can cope with by itself, and the 5V supply is used because the 3V supply cannot provide enough current. Pin D5 is controlled as an analog output to manage the level of vibration with the software, keeping the device as quiet as possible; this also prevents the motor from burning out, as most vibration motors are 3V rather than the 5V the Arduino usually uses.

Note that strictly speaking, the motor should be accompanied by a diode to protect the Arduino from current spikes from the motor, but a little testing with one of these tiny motors showed that a very minimal amount of noise was added to the Arduino supply rails. So for the sake of keeping things simple the normal diode was omitted.

This project uses a protoshield rather than the screwshields used in most of the projects in this book. A protoshield is like a screwshield but without its screw terminals and hence is a bit cheaper and smaller.

STEP 1: ASSEMBLE THE PROTOSHIELD

Protoshields sometimes come with a full set of extra components, such as reset switches and header pins, but for this project you don't want glowing power LEDs that might attract unwanted attention. Therefore, it's better (and cheaper) to buy the bare Protoshield PCB and some headers.

Solder the header pins to the outermost rows of holes on each side of the PCB. A good way to keep the header pins straight is put them into an Arduino and then put the Protoshield PCB on top of the headers. When all the pins are attached, the protoshield should look something like Figure 11-4.

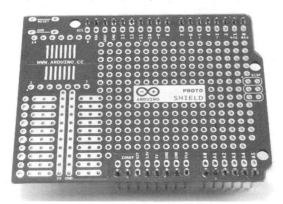

FIGURE 11-4: A PROTOSHIELD WITH HEADER PINS ATTACHED

STEP 2: FIX THE COMPONENTS IN POSITION

Use Figure 11-5 as a reference for the location of the components. All the connections to the NRF24 module are to the 2×4 header on the right of the module's PCB. Don't solder the vibration motor just yet; it will need to be glued in place first as the leads are a bit delicate.

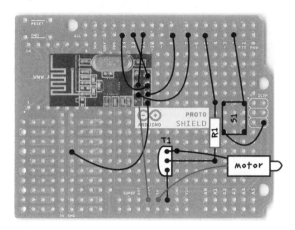

FIGURE 11–5: THE PROTOSHIELD LAYOUT, WHERE R1 IS THE RESISTOR, S1 IS THE SWITCH, T1 IS THE TRANSISTOR, AND THE DARK RECTANGLE AT THE TOP LEFT IS THE NRF24

Apart from the two wires coming from the motor, the dark lines going to various solder pads in Figure 11-5 represent connections you'll make on the underside of the board. The header pins of the NRF24 module fit through the holes in the protoshield, so place that now and solder it to the pads beneath. Do not clip the excess pin lengths off but instead gently splay them out after soldering; this will make it easier to connect them up later. Note that one pin on the NRF24 module is not used.

The transistor has one curved side. It is important that this goes onto the protoshield the right way around, with the curved side pointing left toward the NRF24 (use Figure 11-4 as a guide). Leave about 1/3 inches (about 7.5 mm) of the transistor lead on the top side of the screwshield and fold it down (Figure 11-5) to solder.

The switch has contacts that are on a rectangular grid, four holes long one way and three holes the other. Make sure the switch goes the right way around (Figure 11-4) so that it is longer vertically.

Do not clip off any wires yet, as these can be used to link up the components on the underside of the board. When all the components have been fixed in place, the board should look something like Figure 11-6.

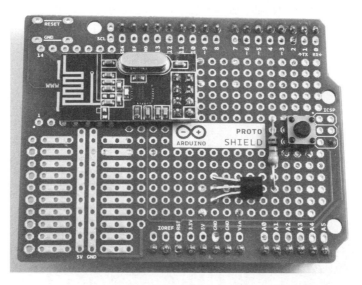

FIGURE 11-6: THE COMPONENTS ATTACHED TO THE PROTOSHIELD

STEP 3: WIRE THE UNDERSIDE OF THE BOARD

This step is the fiddliest, so take care with it. All the components need to be connected on the underside of the board (Figure 11-5). Of course, when the board is flipped over, everything is reversed. In Figure 11-7, I've transposed Figure 11-5 to show the underside of the board for you to work from.

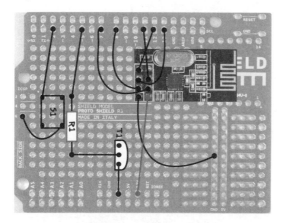

FIGURE 11-7: WIRING DIAGRAM FROM THE UNDERSIDE
OF THE PROTOSHIELD

Figure 11-7 marks the positions of the components so that you can orient yourself, but remember that this is the underside of the board, so the components are actually on the other side of the protoshield.

Many of the connecting wires cross over each other, so use insulated solid-core wire. When everything is connected, the underside of the board should look like Figure 11-8.

FIGURE 11-8: THE UNDERSIDE OF THE PROTOSHIELD

Double-check everything very carefully to make sure there are no accidental solder connections and that every wire makes the correct connection.

STEP 4: ATTACH THE VIBRATION MOTOR

Glue the motor to the *top* of the protoshield, being careful not to get glue on the rotating bit at the front of the motor. The leads are quite fine, so it's better to solder them to the top of the board rather than through a hole. Figure 11-9 shows the motor glued in place and the leads soldered to the protoshield.

FIGURE 11-9: ATTACHING THE VIBRATION MOTOR

STEP 5: REPEAT FOR THE OTHER HANDSET

Having built one handset, do the whole lot again for its partner.

STEP 6: PLACING IT INTO AN ENCLOSURE

You may want to scavenge for some small plastic boxes to contain the handsets. Alternatively, you might prefer to go postapocalypse chic and just tape the battery to the Arduino and protoshield, leaving the battery clip accessible as a rudimentary switch.

SOFTWARE

All the source code for this book is available from *http://www.nostarch.com/zombies/*. See Appendix C for instructions on installing the Arduino sketch for this project, which is called *Project_20_Haptic_Communicator*.

This project uses a community-maintained Arduino library called *Mirf*. This library provides an easy-to-use wrapper around the Serial Peripheral Interface (SPI) serial interface to the NRF24 radio module, allowing the Arduino to communicate with the module. The Mirf library must be downloaded from the Internet, which is another good reason to make this project before the outbreak spreads too far. Download the ZIP file for the library from *http://playground.arduino.cc/InterfacingWithHardware/Nrf24L01*.

Extract the ZIP file and copy the whole *Mirf* folder into *My Documents\ Arduino\Libraries* if you're using Windows or *Documents/Arduino/libraries* if you're using a Mac or Linux. Note that if the *libraries* folder doesn't exist within the Arduino directory, you'll need to create it before copying.

The Arduino IDE won't recognize the new library until you restart it, so after copying the *library* folder, save anything you're working on, quit the IDE, and restart. Next, open the sketch file for this project and upload it to both Arduinos, one after the other. The sketch starts by importing three libraries:

```
#include <SPI.h>
#include <Mirf.h>
#include <MirfHardwareSpiDriver.h>
```

The SPI library is part of the Arduino IDE distribution and simplifies communication with devices using SPI. The MirfHardwareSpiDriver library is also used in the sketch.

Next, three constants are defined:

```
const int numberOfSends = 3;
const int buzzerPin = 5;
const int switchPin = 2;
```

The range of wireless communication can be extended by sending the "button pressed" message several times, so that at the edge of the range, only one of the messages has to get through. The constant `numberOfSends` defines how many times each message should be sent. This is followed by pin definitions for the buzzer and switch pins.

The next constant (`buzzerVolume`) specifies the `analogWrite` value for the vibration motor:

```
const int buzzerVolume = 100; // Keep less than 153 for 3V!
const int buzzMinDuration = 20;
```

If you are using a 3V motor, it is important that the `analogWrite` value does not exceed 153; a value of 153 will deliver power equivalent to a 3V supply to the motor, and more power would overload it. Reducing this value will make your buzzer quieter. The `buzzMinDuration` constant specifies the minimum duration for a buzz in milliseconds. This is important because too short a buzz may not be noticed.

The global `byte data` array contains a 4-byte message to be sent whenever the button is pressed:

```
byte data[] = {0x54, 0x12, 0x01, 0x00};
```

The first three values in this array are chosen as being unique for the pair of haptic communicators. When a message is received, they are checked to see whether they match. This ensures that the communicator has received a real message and not just noise. It also means that you could set up a second pair of devices using different values, and the new pair would not interfere with this pair. Depending on the group dynamics in your band of survivors, you might want to communicate with one person in some situations ("Come save me!") and another person in other situations ("If you show up now, I bet the zombie will eat your brains and not mine").

The fourth byte is not used in this project, but it's there in case you would like the button-press messages to send a parameter. You could, for example, add a second button to the communicator for emergencies that sends a different value in this byte, which could then be read at the receiving end.

Next is the `setup` function:

```
void setup()
{
  analogWrite(buzzerPin, 0);
  pinMode(switchPin, INPUT_PULLUP);
  Mirf.spi = &MirfHardwareSpi;
  Mirf.init();
```

```
  listenMode();
  Mirf.payload = 4;
  Mirf.config();
}
```

This function starts by making sure the buzzer is off at `analogWrite`. Then it sets the mode of the `switchPin` to an input with the internal pull-up resistor enabled (see "Stabilizing Digital Inputs with Pull-up Resistors" on page 252 for more information on pull-up resistors). The radio module is then initialized and put into listen mode, waiting to receive a message.

Next comes the `loop` function:

```
void loop()
{
  if (!Mirf.isSending() && Mirf.dataReady())
  {
    Mirf.getData(data);
    checkForBuzz();
  }
  if (digitalRead(switchPin) == LOW)
  {
        sendBuzz();
  }
}
```

This starts with an `if` statement that first checks whether the module is itself sending a message. It then checks whether there is data ready to be read, and it reads the message over the radio. Once the message is read, the function `checkForBuzz` is called to check that the message is legitimate before buzzing the vibration motor.

The `loop` function finally checks for a button press on this end and responds to a button press by calling the `sendBuzz` function.

Now, let's look at the other functions defined in this sketch, starting with `listenMode` and `sendMode`:

```
void listenMode()
{
  Mirf.setRADDR((byte *)"serv1");
}
void sendMode()
{
  Mirf.setRADDR((byte *)"clie1");
}
```

The `listenMode` function puts the radio module into listening mode by setting its receive address to `"serv1"`. The `sendMode` function puts the radio module into sending mode by setting its receive address to "clie1". We call both the listenMode function and the sendMode function inside sendBuzz, which gets called in the loop function's last if statement.

Finally, we have the checkForBuzz function:

```
void checkForBuzz()
{
  if (data[0]==0x54 && data[1]==0x12 && data[2]==0x01)
  {
    analogWrite(buzzerPin, buzzerVolume);
    delay(buzzMinDuration);
    analogWrite(buzzerPin, 0);
  }
}
```

This function checks the first 3 bytes of the message sent from the other module, and if they match, it turns on the vibration motor for the duration specified in milliseconds by buzzMinDuration.

USING THE HAPTIC COMMUNICATOR

This project is a lot of fun to use. I'm pretty sure casinos are wise to this kind of contraption, though, so to avoid trouble, don't use it to cheat at the gaming tables. Money will have little use after the apocalypse in any case.

If you're prepared to learn Morse code, the handsets can be used with Morse, although they are a little slow. Alternatively, you could come up with a simplified vocabulary along the following lines:

- **One short buzz**: All is well
- **One long buzz**: Zombies sighted
- **Three long buzzes**: Zombies close
- **Three short buzzes**: Run!!

This is the final project in the book, and I hope you have had fun as you've equipped yourself for the apocalypse. Whether you're building these projects in anticipation of the coming zombie hordes or you're already in hiding, I also hope they help you to survive!

A

PARTS

In this appendix, you will find more information about the parts used to make the projects in this book. Unlike the individual project supply lists, the tables in this appendix list two types of sources: preapocalypse and postapocalypse. If you're looking to buy parts and stockpile them in your secret underground bunker before the dead rise, look to the suppliers in the preapocalypse column. Buy your materials in bulk online now, and you can even order extras so you'll be prepared to replace any components that break.

If you're reading this guide after zombies have already taken up residence in your neighborhood, you want the postapocalypse column. Your options will be limited without the Internet, but if you're lucky, you'll find the odd brick-and-mortar hobby shop to loot, and there should be plenty of cars, microwaves, and other electronics that you can harvest components from. Good luck!

A NOTE ON BRICK-AND-MORTAR SUPPLIERS

When it comes to brick-and-mortar stores for electronic components, since the demise of Radio Shack, your choice in the United States has been reduced pretty much to Fry's Electronics in California, Texas, and a handful of other states (*http://www.frys.com/*) and a few independent stores around the country. If you live in the UK, then Maplin Electronics (*http://www.maplin.co.uk/*) is your best bet. Both Fry's and Maplin offer online ordering as well.

ELECTRONICS MODULES

This section describes items that could loosely be termed modules, or preassembled parts, rather than basic electronic components.

ITEM	PREAPOCALYPSE SOURCE	POSTAPOCALYPSE SOURCE
7A (or more) 12V charge controller	eBay, Fry's (4980091)	Abandoned RVs and boats
Arduino Uno R3	Adafruit, Fry's (7224833), SparkFun	Fry's
screwshield	Adafruit (196)	
LCD shield	eBay, SparkFun (DEV-11851)	
PIR module	Adafruit (189), Fry's (6726705)	Fry's, security store
Door latch	Farnell	Fry's, security store
RF remote single-channel relay, 12V	eBay	
Reed switch and magnet pair	Adafruit (375), Fry's (1908354)	Fry's, security store
4-channel relay shield	eBay, http://www.sainsmart.com/	
USB Bluetooth adapter	eBay	computer store
HC-06 Bluetooth serial module	eBay	
servo motor (small, 9 g)	Adafruit (196), eBay	Hobby store
servo motor (standard)	Adafruit (155), eBay	Hobby store
NRF24 radio module	eBay	
Protoshield	eBay (Arduino code: A000077)	

RASPBERRY PI AND RELATED PARTS

This list includes all Raspberry Pi–specific parts you'll need, including the Pi itself.

ITEM	PREAPOCALYPSE SOURCE	POSTAPOCALYPSE SOURCE
Raspberry Pi	Adafruit (2358), Fry's (8258726)	
Small HDMI monitor	Adafruit (1934), eBay	
Raspberry Squid	Amazon, http://www.monkmakes.com/	

LEADS AND CONNECTORS

In this list, you'll find all the wires, leads, jacks, and other bits you'll need to connect your circuits.

ITEM	PREAPOCALYPSE SOURCE	POSTAPOCALYPSE SOURCE
Heavy-duty alligator-clip leads (7A or more)	Auto parts store	Auto parts store
Terminal block (10A)	Home Depot, Lowe's, Menards	Home Depot, Lowe's, Menards
Small alligator clip leads	Auto parts store	
Terminal block (2A)	Home Depot, Lowe's, Menards	Home Depot, Lowe's, Menards
Female-to-female jumper wire	Adafruit (266)	
0.1 inch header pins	Adafruit (392), eBay	
Female-to-male jumper wire	Adafruit (826)	
2.1 mm jack plug-to-cigarette lighter adapter	Auto parts store	Auto parts store
2.1 mm barrel jack with flying leads	Broken DC power supply	DC power supply
Long male-to-male jumper wires (20 cm)	Adafruit (760)	
0.1 inch right-angle header pins	eBay	
9V Arduino battery lead	Adafruit (80), eBay	
Solid-core wire for proto-screwshield PCB links	Adafruit (1311)	Abandoned electronics

TOOLS

No self-respecting zombie apocalypse survivor should be without the following general household tools:

- A drill
- Screwdrivers
- Pliers
- Snips
- A wood saw
- Scissors

You should be able to find these at any hardware store. To complete the projects in this book, you will also need a few electronics construction tools, listed below.

ITEM	PREAPOCALYPSE SOURCE	POSTAPOCALYPSE SOURCE
Multimeter	Auto parts store, eBay, Fry's	Auto parts store, Fry's
Soldering iron	Auto parts store, Fry's	Auto parts store, Fry's

ELECTRONIC COMPONENTS

A lot of the components here can be found in electronics starter kits for hobbyists. Kits like Adafruit's ARDX Experimenters Kit for Arduino (product ID 170) or the SparkFun Beginners Parts Kit (KIT-10003) will give you a good start with the basic resistors, diodes, and transistors.

ITEM	PREAPOCALYPSE SOURCE	POSTAPOCALYPSE SOURCE
Piezo buzzer	Adafruit (1740), eBay	
270 Ω resistor	Mouser (293-270-RC)	
470 Ω resistor	Mouser (293-470-RC)	
Push button	Adafruit (1439)	
1 kΩ resistor	Mouser (293-1k-RC)	
1N4001 diode	Adafruit (755)	
Blue or white LED	Adafruit (301)	
100 µF ceramic capacitor	Adafruit (753)	
TMP36	Adafruit (165)	
Microswitch	Fry's (2314449)	Microwave oven

ITEM	PREAPOCALYPSE SOURCE	POSTAPOCALYPSE SOURCE
Small sealed lead acid battery	Fry's (6607854), security store	
FQP33N10 or FQP30N06 MOSFET	Adafruit (355)	
Resistor (100 Ω 2W)	Mouser (594-5083NW100R0J)	
Resistor (100 Ω 1/4W)	Mouser (293-100-RC)	
High-volume buzzer	security store	security store, smoke alarm
2N3904 NPN bipolar transistor	Adafruit (756)	
5V or 3V vibration motor	eBay	
Tactile push switch	Adafruit (504)	
Red LED	Adafruit (297)	

OTHER HARDWARE

Finally, you'll need just a few other odds and ends to be able to power and construct the mechanics of your projects, as listed here.

ITEM	PREAPOCALYPSE SOURCE	POSTAPOCALYPSE SOURCE
A100 V drive belt	Auto parts store, eBay	Auto parts store, hardware store, scavenge
Project box	Fry's	closets, garages
4 × AA battery box	Adafruit (830)	
6 × AA battery box	Adafruit (248)	

RESISTOR COLOR CODES

Resistors have stripes on them that tell you their value, and an essential piece of geekiness is to know your resistor color codes.

COLOR	VALUE	COLOR	VALUE
Black	0	Blue	6
Brown	1	Violet	7
Red	2	Gray	8
Orange	3	White	9
Yellow	4	Gold	1/10
Green	5	Silver	1/100

There will generally be three of these bands together starting at one end of the resistor, a gap, and then a single band at the other end of the resistor. The single band indicates the accuracy of the resistor value. While gold and silver represent the fractions 1/10 and 1/100, they're also used to indicate how accurate the resistor is; gold is ±5 percent and silver is ±10 percent.

Figure A-1 shows the arrangement of the colored bands. The resistor value uses just the three bands. The first band is the first digit, the second the second digit, and the third "multiplier" band is how many zeros to put after the first two digits.

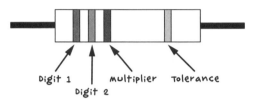

FIGURE A-1: RESISTOR COLOR CODES

Let's say the digit-1 band is red, the digit-2 band is violet, and the multiplier band is brown. That makes this a 270 Ω resistor, or 27×10¹. Similarly, a 10 kΩ resistor will have bands of brown, black, orange (1, 0, 1,000).

B

BASIC SKILLS

If you're going to be a postapocalyptic maker and survive the land of the walking dead, then you'll need a few key electronics skills. This appendix is a quick guide to the basics, such as joining wires together, soldering, and using a multimeter. Flip here anytime you need a refresher. It may save your life!

STRIPPING WIRES

For an apocalypse survivor, stripping the insulation off wires is a skill that belongs near the top of the list. The devices in this book will help you stay alive, and to build them, you'll often need to join insulated wires together or fit them into a screw terminal. The first step in that process is exposing the bare wire.

To strip a wire, use a blunt pair of pliers to grip the wire and pull off the insulation with a sharp pair of wire cutters (also called snips). Figure B-1 shows the process.

A

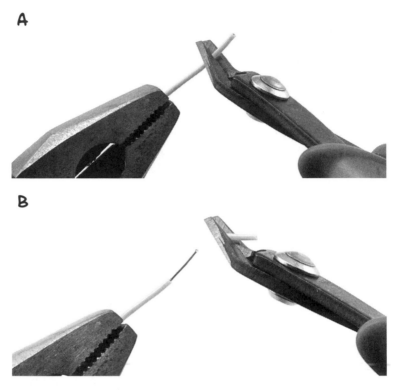

B

FIGURE B-1: STRIPPING WIRES

Grip the wire with pliers (Figure B-1a). If your wire is long, you could wrap it around your fingers instead. Either way, the idea is to stop the wire from moving. Next, gently pinch the wire with the cutters at the position where you want to remove the insulation. Apply just enough pressure to almost cut through the insulation without cutting into the wire inside, then pull the insulation away (Figure B-1b). If the snips start to slip as you pull, just squeeze them a bit tighter.

Mastering this skill can take a while, so practice on some old wire before you try it on something important. If you cut the last good wire in your cache too short, you could find yourself unable to complete your latest antizombie invention until the next supply run—when it might be too late.

JOINING WIRES BY TWISTING

Knowing how to twist wires together is a useful skill, too, especially if you haven't come across any solder in your scavenging trips. If done properly (as illustrated in Figure B-2), just twisting the wires together can make pretty good electrical connections.

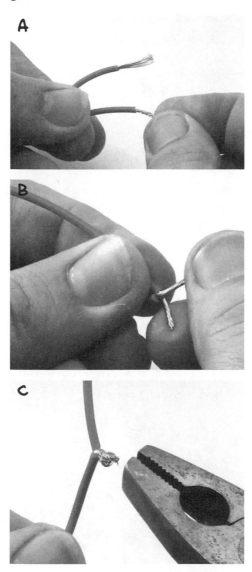

FIGURE B-2: JOINING WIRES BY TWISTING

First, strip about half an inch (15 mm) of insulation off each wire (see "Stripping Wires" on page 227). Then, if your wire is stranded rather than solid, use your thumb and forefinger to twist each wire on its own and keep all the strands together (Figure B-2a). Next, place the two wires side by side, lining up the ends of the insulation, and twist the wires around each other (Figure B-2b). Try to make sure that the wires actually go around each other, rather than leaving one wire straight while the other wraps around it. This can be difficult if the wires are of different thicknesses.

Finally, coil the intertwined wires into a tight ball (Figure B-2c) and wrap the whole thing in electrical tape or heatshrink tubing (see "Using Heatshrink" on page 235). You can also use pliers to really tighten up the joint.

If you have soldering equipment, then you can make the connection mechanically stronger and more electrically reliable by heating the little knot with a soldering iron and feeding solder into it, as I describe in the next section.

If you want to know how NASA does it, take a look at this link: *http://makezine.com/2012/02/28/how-to-splice-wire-to-nasa-standards/*.

SOLDERING BASICS

Soldering is much easier than it looks, and you don't need to spend a lot of money on a fancy soldering station. During an apocalypse, your options will be limited, but a basic starter kit (see Figure B-3) will work just fine.

You can find basic soldering kits at an auto parts store or even at some hardware stores. If you are buying in advance of the apocalypse, then Adafruit sells a great starter kit (product 136) that also includes a multimeter, hookup wire, and various other useful bits and bobs.

FIGURE B-3: A BASIC SOLDERING KIT

There are lots of accessories and tools that can make soldering quicker, but these are by no means essential. Here's all you really need:

A soldering iron   Look for an iron with a power rating of 30W or more, with a fine tip (say 1/25 inch, or 1 mm). Before the zombie apocalypse, just buy one that's AC powered. To prepare, you could also buy a soldering iron that runs on 12V DC and keep it with your emergency supplies; that way, you'll have an iron you can power from a car battery. These soldering irons, intended for working on the electrical components of cars, are quite common.

Solder   If you buy a soldering kit, it will probably come with a coil of solder. Solder comes in two flavors: leaded and lead-free. Leaded solder melts at a lower temperature and is generally easier to use than lead-free solder. But please don't eat either, no matter how desperate your food situation becomes.

Snips   You'll need a good pair of wire cutters to cut wires close to the surface of a PCB and for stripping wire.

A damp sponge or cloth   Any old sponge will do. You'll use it to wipe the tip of the iron when there's excess solder.

WARNING Soldering irons get hot. In fact, they get really hot, much hotter than the maximum temperature of your kitchen oven. So it goes without saying that if you touch the hot end of a soldering iron, you'll get a serious burn. This is not an activity for unsupervised children. Similarly, lead is a toxic element that is not at all good for you, so you may prefer to use lead-free solder, despite it being a little harder to work with.

JOINING WIRES WITH SOLDER

To join together two wires with solder, start by following the instructions in "Joining Wires by Twisting" on page 229. Then, you can solder the joint. The trick with soldering is to always make the solder flow into the thing you're soldering; Figure B-4 shows solder flowing into the ball of wires from Figure B-2c.

Many beginners make the mistake of creating a blob of solder on the tip of the iron and then blobbing it onto the wire. This usually results in poor quality *dry joints* that may look okay but will fail quickly and, before they fall apart, may not make good contact with the wire. Therefore, you'll want to heat the wire you want to solder before you touch the solder to it.

FIGURE B-4: RUNNING SOLDER INTO THE JOINED WIRES

With that in mind, you can join your twisted wires as follows:

1. Turn your iron on and leave it to heat up. If your kit didn't come with a stand for the soldering iron, make sure that you prop it somewhere safe so that the hot end is not touching anything.

2. Touch the end of the solder to the tip of the iron to see if it is hot. If it immediately melts and flows over the tip of the iron, then the iron is ready.

3. If the tip of the iron is not shiny and bright after this, then wipe it on a wet sponge. This makes a great sizzling noise! Repeat the previous step to *tin* the tip of the iron with solder. *Tinning* just means coating a wire or the tip of your iron with solder by heating it up and then pushing the solder onto it.

4. Press the tip of the soldering iron against the little knot of wires and leave it there for perhaps three or four seconds. Then, with the soldering iron still pressed to the knot, push the end of the solder onto the knot. The solder should flow into the knot. If your solder isn't flowing well, it sometimes helps to feed a bit more solder to the joint, as solder contains cores of *rosin flux*, which helps the solder to liquefy.

5. Keep feeding the solder in until the whole knot of wires is coated in solder.

6. Remove the tip of your solder thread from the joint and replace the iron on its stand. Make sure that the wires don't move while you give them 10 or 20 seconds to cool down.

You can also insulate your soldered connection with electrical tape or with heatshrink, as described in "Using Heatshrink" on page 235. If you plan to do this, then you can make a neater joint by soldering the wires side to side, without twisting them together (Figure B-5a–e).

After stripping the ends of the wires (Figure B-5a), tin them with solder (Figure B-5b). If the wire is stranded, the solder should flow between the strands that make up the wire.

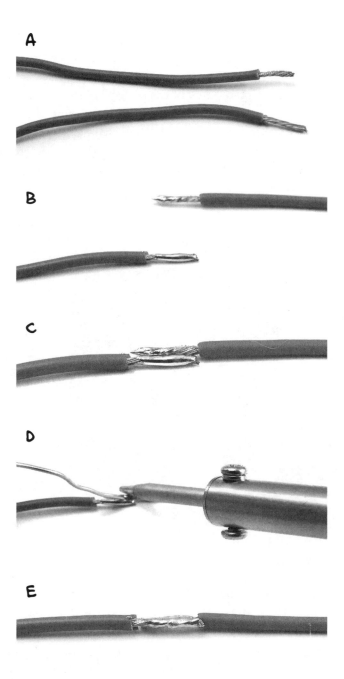

FIGURE B–5: SOLDERING WIRES TOGETHER WITHOUT TWISTING THEM FIRST

Now, lay the wires next to each other (Figure B-5c), heat the wires, and run solder into the valley they make (Figure B-5d). The end result should be a nice, even joined pair of wires (Figure B-5e).

SOLDERING A PCB

Wires are easier to scavenge than complete circuits, but being able to solder to a printed circuit board (PCB) will certainly serve you well during an apocalypse. For example, quite a few of the projects in this book use a screwshield that requires a bit of soldering to put together. Fortunately, the screwshield is a PCB with lots of convenient metal pads that are made for soldering. If you successfully followed the steps described in "Joining Wires with Solder" on page 231, then you shouldn't have any problems soldering a PCB.

When attaching components to a PCB, the basic idea is that you push a component's legs through from the top, flip the PCB over, solder the leads to their solder pads, and snip off the excess wire. Figure B-6 shows a component lead being soldered onto a screwshield.

FIGURE B-6: SOLDERING A COMPONENT LEG TO A PCB

As with all soldering, the trick is to apply the solder to the thing being heated up rather than to the soldering iron, so heat the component leg and touch solder to it. You'll often get the best results by giving the soldering iron a second or two to heat the component lead and solder pad before you apply the solder to the junction of the soldering iron tip and the component lead. Figure B-7 shows examples of two solder joints, one bad and one good.

The solder joint on the left is best described as "blobby," and it's a result of allowing a glob of solder to form on the tip of the iron and then "blobbing" it onto the PCB. The solder joint on the right is close to perfect. See how the whole pad is covered in solder, flowing all the way around the component lead and forming a very gentle little hill of a meniscus.

FIGURE B-7: BAD (LEFT) AND GOOD (RIGHT)
SOLDER JOINTS

USING HEATSHRINK

When you're confident in your wire-connecting skills, try using heatshrink to insulate the wires. Heatshrink is a great way to finish a pair of wires that have been joined by twisting or soldering, and it's a lot more durable than electrical tape. Wrapping the wires in electrical tape is fine at first, but eventually the tape starts to lose its stickiness and unravel. Heatshrink is also just more fun to use, and you'll need all the fun you can get when zombies are the only ones knocking on your door.

Heatshrink comes as a tube that you can cut to the length you need. When heated with a hair dryer, hot air gun, or even a cigarette lighter, it shrinks to about half its diameter as if by magic. If your heatshrink starts out with a fairly snug fit over the wires, then it will grip the wires tightly after you heat it.

Here's how to make a good connection and strengthen it with heatshrink:

1. Choose a heatshrink tube slightly wider than the joint you want to cover. Cut a sleeve long enough to cover the exposed wire and overhang onto the wire's insulation a little bit.

2. If you're connecting two wires that already have parts attached to their other ends, slide your heatshrink sleeve onto one wire before you solder them together, pushing the heatshrink as far away from the solder point as possible. I've lost count of the number of times I have soldered something together only to remember too late that the heatshrink then couldn't be slid on. Every time that happens, I have to unsolder the wires again.

3. Join the wires using the end-to-end method described in "Joining Wires with Solder" on page 231. You'll end up with something like Figure B-8a.

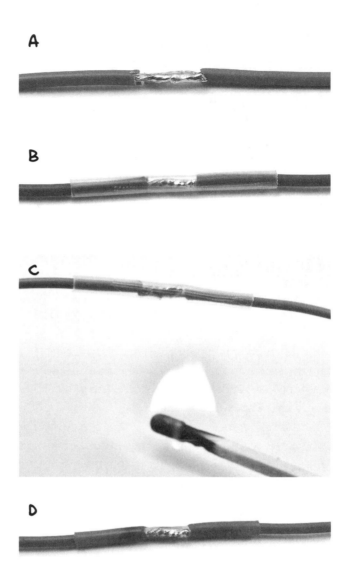

FIGURE B–8: APPLYING CLEAR HEATSHRINK TUBING OVER JOINED WIRES

4. If you haven't already done so, slide the heatshrink sleeve over the joint (Figure B-8b). The heatshrink I show is clear so you can see that the solder joint is good. Heatshrink is also commonly available in black and other colors.

5. Heat up the heatshrink with a hair dryer or even a match held underneath it (Figure B-8c). You don't need to make it super hot. Just keep heating until you have a nice tight fit, as in Figure B-8d. But try not to scorch it!

Heatshrink comes in a huge range of diameters. If you plan to use it, I suggest buying a selection box that has short lengths of various diameters of heatshrink tubing. You can find these at auto parts stores, as heatshrink is often used when modifying or repairing car wiring.

USING A MULTIMETER

An electric current is a flow of electrons. But electrons are small—very small, in fact. So when it comes to working out what's going on electrically, we need something that will allow us to measure what those pesky electrons are up to.

Where a doctor has a stethoscope to check the various pulse points in your body, an electronics enthusiast will use a multimeter (Figure B-9) to check specific points on a circuit.

FIGURE B-9: A MULTIMETER

The multimeter shown in Figure B-9 cost about $5 but is still more accurate and has a wider range of features than an expensive multimeter from 20 years ago. Something like it should be perfectly good for any current, voltage, or resistance you need to measure to get ready for the zombie apocalypse.

A multimeter consists of a display at the top, a big rotary switch in the middle to select different measurement ranges, and some sockets at the bottom for attaching test leads. A multimeter should include test leads when you buy it. These are usually of the sort shown in Figure B-10a, but it can be very useful to also get some test leads that have alligator clips on the end (Figure B-10b).

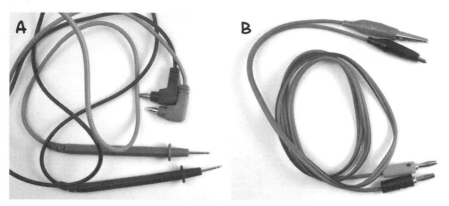

FIGURE B-10: TEST LEADS

Most auto parts stores will have multimeters, and many places where you can buy tools might well have a multimeter or two. Amazon and eBay also have a huge array of low-cost multimeters for you to choose from, if you want to stock a couple in your apocalypse preparedness kit.

MEASURING DC VOLTAGE

Multimeters are most commonly used to measure DC voltage. This is what we would do to, say, check the voltage of a battery (Figure B-11).

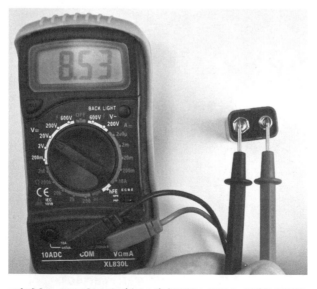

FIGURE B-11: MEASURING DC VOLTAGE WITH A MULTIMETER

If the battery says on the case that it's a 9V battery, but when you measure the voltage across its terminals you get a reading of 4V, then there is something wrong with the battery. The 9V battery in Figure B-11 measures 8.53V, which is perfectly normal. If it's under 8V, you should probably toss it.

To measure the voltage of a battery, follow these steps:

1. Set the range knob of the multimeter to DC volts and pick a range that is higher than the highest voltage you are expecting. For a 9V battery, for example, the 20V range is a good choice. (Multimeters also have an AC voltage range. The AC ranges have a wavy line next to them, and the DC ranges have one horizontal line above another.)

2. Make sure that the test leads are in the sockets for voltage measurement and *not* for current measurement. The black lead should be plugged into the COM socket, and the red lead should be plugged into the socket marked with a V. This is important because when measuring current, the multimeter leads are almost a short circuit and using a current-configured meter to measure voltage would cause a short circuit across the battery. This is likely to blow a fuse in the multimeter.

3. Connect the black COM lead to the negative end of the battery and the red positive lead to the positive terminal of the battery. The multimeter's display will tell you the voltage.

In addition to measuring the voltage of a battery to find out whether it's good, you may want to measure the voltage across a component, say an LED or resistor. In that case, just touch the probe leads to either side of the component.

MEASURING DC CURRENT

When you need to maximize the life of your battery, which will be important when the apocalypse is on, it's often useful to see how much current a device is using. As an example, we could test how much current will be drawn by an Arduino.

Figure B-12 shows a multimeter set up to test the current consumption of an Arduino powered from a 9V PP3 battery. A barrel jack lead is used to connect the 9V battery. The multimeter sits in the circuit, measuring the current flowing through it (in this case 32.6 mA). The positive terminal of the battery is connected to the positive lead of the multimeter, and the rest of the circuit (or in this case the Arduino) receives its power through the negative lead of the multimeter.

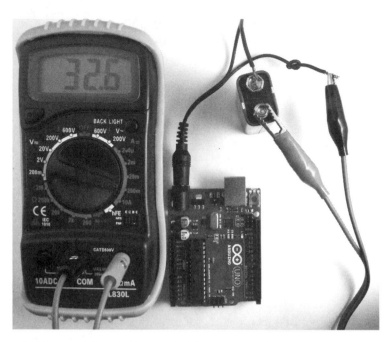

FIGURE B-12: MEASURING DC CURRENT WITH A MULTIMETER

Follow these steps to measure current:

1. Set the range knob of the multimeter to a DC Amps range. On its own, an Arduino only uses about 30mA of current, so select the 200mA range. If in doubt, start with the maximum range (often 10A) and work down if you need more precision.

2. Make sure that the positive test lead is in the correct current measuring socket on the multimeter. For low currents (about 200mA or less), this is often the same connection that's used to measure volts. The multimeter shown here has a separate socket for currents up to 10A, but since we shouldn't see more than 30mA, the voltage socket is being used.

3. Connect the positive test lead of the meter to the positive side of the battery and the negative test lead to the positive voltage connection of the lead to the Arduino.

As shown, the multimeter is effectively intercepting the current flowing through the test leads in order to measure the current.

MEASURING RESISTANCE

"Resistor Color Codes" on page 225 includes a guide to identifying the values of resistors from their color stripes. Another way to find the value of a resistor is to measure it using a multimeter. Just set the meter to one of its resistance ranges and then touch the two test leads to either side of the resistor (Figure B-13).

FIGURE B-13: MEASURING RESISTANCE WITH A MULTIMETER

In this case, the resistor is measured as 118.2 Ω. The resistor's nominal value, according to the stripes, is 120 Ω. This slight discrepancy is perfectly normal. Neither the multimeter nor the resistor itself will be completely accurate.

NOTE Some meters also have one or more capacitance ranges, which you can use to measure the value of capacitors in the same way.

CONTINUITY TESTING

Most multimeters have a Continuity or Buzzer mode, selectable from the range knob. When the multimeter is set to continuity, a buzzer on the multimeter sounds if the two test leads are touched together. The buzzer should also sound when the leads are connected by something with low resistance, like a wire, PCB track, or dubious solder joint.

This function may not sound very useful, but it is actually invaluable. It allows you to test fuses as well as suspect wires that look okay but may have a break beneath the insulation. It is also good for testing switches. Just touch the leads the switch contacts, and if the multimeter buzzes when you flip the switch, then all is well. Similarly, to test a fuse, first touch the test leads together to hear the beep and make sure the multimeter is working and then touch the leads to either end of the fuse. If the meter doesn't work, then the fuse has blown.

BELLS AND WHISTLES

The multimeter features I've already described will cover pretty much any test you might need to perform on a circuit in this book. However, even a cheap multimeter, like the one shown here, has some other useful settings:

AC voltage and current   A separate set of ranges are needed for AC because it swings both positive and negative, making its average value zero, so the meter will convert the AC to DC internally before giving a reading if one of these ranges is selected.

HFE   This range will measure the gain (current amplification factor) of a transistor plugged into the special transistor socket. This is also a quick way to see whether a transistor is dead.

If you buy a more expensive multimeter, you will find it has even more bells and whistles:

Frequency measurement   Measures the frequency of a signal. You could, for example, use this to find the frequency of the buzzer on the smoke alarm in "Project 11: Quiet Fire Alarm" on page 120.

Temperature   This function requires a special thermocouple probe. It's useful as a general thermometer and is especially valuable as a way to see if components are getting dangerously hot.

Capacitance   This setting is useful for comparing the capacitance written on the side of a capacitor with its actual capacitance. Electrolytic capacitors are notoriously unreliable as they get older. They often degrade into a zombie-like state, causing problems in many kinds of electronic equipment.

Backlight   Lights the screen on your multimeter, which is useful if you are trying to use the multimeter to work out why the lights in your base have gone out!

Auto power off   Very handy if, like me, you tend to forget to switch things off. You never know when you'll find more batteries, after all.

Your multimeter will be one of your most useful tools, so get familiar with it. That way, should you have to use it under pressure as the zombies close in, you won't have to waste valuable time consulting the manual.

C

ARDUINO PRIMER

 Arduino microcontroller boards are perfectly suited to a postapocalyptic world. They're robust, they're reliable, and they use very little power. If you're new to Arduino, this appendix will get you started with this great little board so you can begin to make your end-of-the-world preparations now and greatly enhance your chances of survival.

WHAT IS AN ARDUINO?

There are various types of Arduino board, but by far the most common is the Arduino Uno, and this is the one used for all the projects in this book (see Figure C-1).

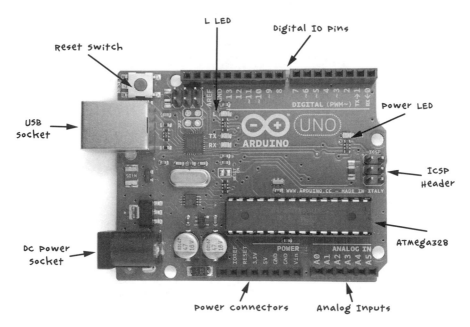

Reset switch

L LED

Digital IO pins

USB socket

Power LED

ICSP Header

DC power socket

ATMega328

Power connectors

Analog Inputs

FIGURE C-1: AN ARDUINO UNO R3

The Arduino Uno shown in Figure C-1 is a revision 3 (R3) board, which is the latest at the time of writing. We'll have a look at each of the components and their uses.

Let's start our tour with the USB socket. This serves several purposes: it can be used to provide power to the Arduino or to connect the Arduino to your computer for programming. It can also serve as a communications link to other computers, as in "Project 13: A Raspberry Pi Control Center" on page 140 where it sends data from the Arduino to a Raspberry Pi. The little red button on the Arduino is the Reset button. Pressing it will cause the program that is installed on the Arduino to restart.

The connection sockets along both the top and bottom edges of the Arduino are where you attach electronics. On the top side of Figure C-1 are digital input and output pins, numbered 0 to 13 and configurable as either inputs or outputs. Inputs read messages coming in; for example, if you connect a switch to a digital input, the input will detect whether the switch is pressed. Outputs send information or power out; if you connect an LED to a digital output, you can turn it on by switching the output from *low* to *high*. In fact, one LED, called the *L* LED, is built onto the board and connected to digital pin 13.

On the right, the power LED indicates whether the board is powered. The ICSP (In-Circuit Serial Programming) header is only for advanced programming of the Arduino, and most casual users of Arduino will never use it.

The ATMega328 is a microcontroller integrated circuit (IC) and the brains of the Arduino. The chip contains 32KB of flash memory, where you store the program you want the Arduino to run.

On the bottom right of Figure C-1 is a row of analog input pins labeled A0 to A5. Digital inputs can only tell whether something is on or off, but analog inputs can actually measure the voltage at the pin, as long as the voltage is between 0V and 5V. Analog input pins could be used, for example, to measure voltage from a temperature sensor like the one used in "Project 12: Temperature Alarm" on page 131.

The final row of sockets provides miscellaneous power connections. In "Project 4: Battery Monitor" on page 53, we use V_{in} (volts in) to provide power to the Arduino; 5V and GND (or ground), which means 0V, are also power connections that you will need when connecting external electronics.

At the bottom left, we have a DC power jack, which is another power connection. This can accept anything between 7V and 12V DC. The Arduino will automatically accept power from the USB socket and power from the DC connector or V_{in} socket, too.

ARDUINO SOFTWARE

The Arduino might not be what you would expect from a computer. It has no operating system and no keyboard, monitor, or mouse. This is, of course, good news for the survivor who needs to travel light. And while you can reprogram an Arduino as many times as you like, it also only ever runs a single program (called a *sketch*) at a time. To program the Arduino, you must have the Arduino IDE software installed on your normal computer, so we'll first cover installation and then talk about writing programs.

INSTALLING THE ARDUINO IDE

The Arduino IDE is easy to use, making it one major reason for the Arduino's great popularity. It is available for Windows, Mac, and Linux computers, and it programs the Arduino over a USB connection without any need for special programming hardware.

NOTE   *You will need an Internet connection to download the Arduino IDE, so do this before you start hearing about zombies on the news!*

To install the Arduino IDE for your platform, download the software from the Arduino site at *http://www.arduino.cc/* (click **Download** at the top and install the version that's appropriate for your system). Then follow the instructions from the Getting Started link. Windows and Mac users will need to install USB drivers for the Arduino IDE to be able to communicate with the Arduino.

Once you have everything installed, run the Arduino IDE. Figure C-2 shows the Arduino IDE window with some code in it.

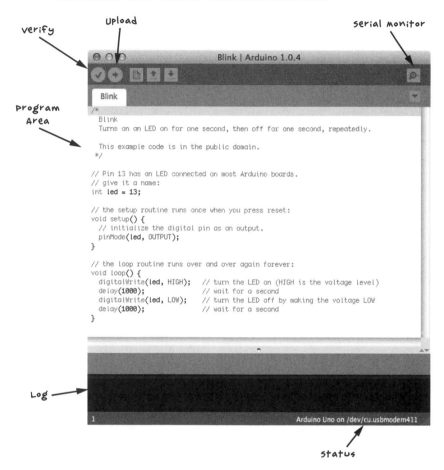

FIGURE C-2: THE ARDUINO IDE WINDOW

The Upload button, as the name suggests, uploads the current sketch to the Arduino board. Before uploading, however, it converts the textual programming code into executable code for the Arduino and displays any errors in the Log area. The Verify button checks the code for errors without uploading the program to the board.

The serial monitor button opens the serial monitor window, which is used for two-way communication between the Arduino and another computer, as in "Project 13: A Raspberry Pi Control Center" on page 140. You can type in text messages to send to the Arduino, and you should see any responses that come back in the same window. The Status area at the bottom of the screen gives information on the type of Arduino you're using and the corresponding serial port that will be programmed when the Upload button is pressed. The Status area in Figure C-2 also shows the type of port you would expect to see when using a Mac or Linux computer (something like /dev/cu.usbmodem411). If you're using a Windows computer, this will display COM followed by a number.

The large, white area of the IDE is the Program Area, where you type the program code you want uploaded to the Arduino.

The File menu allows you to Open and Save sketches as you would in a word processor, and it has an Examples submenu from which you can load example sketches.

UPLOADING A SKETCH

To test out your Arduino board and make sure the Arduino IDE is properly installed, click **File ▸ Examples ▸ 01. Basics** to open the example sketch called *Blink* (shown in Figure C-2).

Use a USB cable to attach your Arduino to your computer. The power LED of the Arduino should light up as it's plugged in, and a few other LEDs should flicker as well.

Now that the Arduino is connected, you need to tell the IDE the type of board being programmed and the serial port it's connected to. Set the board using the menu **Tools ▸ Board** and then select Arduino Uno from the list of boards.

Set the serial port using the menu **Tools ▸ Port**. If you're using a Windows computer, you probably won't have many options there; you may find only the option COM4. On a Mac or Linux computer, there are generally more serial connections listed, many of which are internal devices, and it can be difficult to work out which one refers to your Arduino board.

Usually, the correct port is one that starts dev/ttyusbmodem*NNNN*, where *NNNN* is a number. In Figure C-3, the Arduino attached to my Mac has been selected.

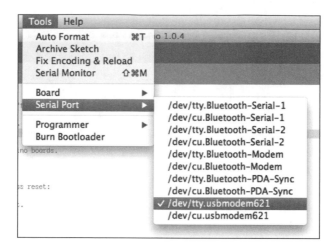

FIGURE C-3: SELECTING THE ARDUINO SERIAL PORT

If your Arduino doesn't show up in the list, this usually means you have a problem with the USB drivers, so try reinstalling them. If you're a Windows user, try rebooting.

You should now be ready to upload the sketch to the Arduino, so press the **Upload** button. Messages should appear in Log area, and then the TX and RX LEDs on the Arduino should flicker as the program is uploaded onto the board.

When the upload is complete, you should see a message like the one shown in Figure C-4.

FIGURE C-4: A SUCCESSFUL UPLOAD

The Done uploading message tells you that the sketch has uploaded, and the last line in the console tells you that you've used 1,084 bytes of the 32,256 bytes available on your Arduino.

Once the sketch is uploaded, the built-in *L* LED on the Arduino should blink slowly on and off, which is just what the *Blink* program is expected to do.

INSTALLING THE ANTIZOMBIE SKETCHES

All the sketches for the book are available via the book's website (*http://www .nostarch.com/zombies/*). Click on the Download Code link to download a ZIP file called *zombies-master.zip*. Make sure to do this *before* the apocalypse begins, because your broadband is likely to be a low priority once the infection has begun to spread. This folder will contain all the Arduino and Raspberry Pi programs for the projects in this book.

Install the Arduino sketches so that you can use them directly from your Arduino IDE by copying the subfolders from the *Arduino* folder into *Documents/Arduino* folder for Mac and Linux users and *My Documents\Arduino* for Windows users. Exit and reopen the Arduino IDE. Now when you view **File ▸ Sketchbook**, you should find all the book's sketches listed.

ARDUINO PROGRAMMING BASICS

This section contains an overview of the main Arduino programming commands to help you understand the sketches used to do with zombies. If you're interested in learning the Arduino C programming language, consider getting a copy of my book *Programming Arduino: Getting Started with Sketches* (Tab Books, 2012). The technical reviewer for the book you're reading now (Jeremy Blum) has also written a very good book on Arduino and has produced a superb series of video tutorials. You can find links to all this from his website (*http://www.jeremyblum.com/*).

STRUCTURE OF AN ARDUINO SKETCH

All Arduino sketches must have two basic *functions* (units of program code that perform a task): setup and loop. To see how they work, let's dissect the *Blink* example that we looked at earlier.

```
int led = 13;

// the setup routine runs once when you press reset
void setup() {

  // initialize the digital pin as an output
  pinMode(led, OUTPUT);
}

// the loop routine runs over and over again forever
void loop() {
  digitalWrite(led, HIGH);   // turn the LED on (HIGH is the voltage level)
  delay(1000);               // wait for a second
  digitalWrite(led, LOW);    // turn the LED off by making the voltage LOW
  delay(1000);               // wait for a second
}
```

Your *Blink* sketch might be slightly different if you have a newer version of the Arduino IDE, so for the purposes of this discussion, refer to the sketch printed here rather than the one loaded in your IDE.

The text preceded by a double slash (//) is called a *comment*. It's not executable program code but rather a description of what's happening at that point in the sketch.

Just after the words setup() and loop(), we have a { symbol. (Sometimes this is put on the same line as the preceding word and sometimes on the next line. Where it goes is just a matter of personal preference and has no effect on the running of the code.) The { symbol marks the start of a block of code, which ends with a corresponding } symbol. You'll use curly brackets to group together all lines of code that belong to a particular function or other control structure.

The lines of code inside the setup function run just once, when power is applied to the Arduino or the Reset button is pressed. You use setup to perform all the tasks that need doing just once when the program starts. In *Blink*, the code inside the setup function just sets the LED pin as an output.

The commands inside the loop function will be run over and over again; in other words, when the last line inside loop has run, the first line will start again.

Now, let's parse this sketch, starting from the top line.

CREATING VARIABLES AND CONSTANTS

Variables are a way of giving names to values; for example, the first line of *Blink* labels pin 13 led:

```
int led = 13;
```

This defines an int variable called led and gives it an initial value of 13, because 13 is the number of the Arduino pin that the *L* LED is connected to. The word int is short for integer and means that this variable returns a whole number without decimals.

In some of the book's other sketches, variables like this, that define a specific pin to be used, are preceded by a const keyword:

```
const int led = 13;
```

The const keyword tells the Arduino IDE that the value of led is never going to change from 13, making it a *constant*. Assigning values this way results in slightly smaller and quicker sketches and is generally considered a good habit.

CONFIGURING DIGITAL OUTPUTS

The Blink sketch also shows a good example of a setting a pin up to be a *digital output*. Pin 13, having been defined as led, is configured as an output in the setup function by this line:

```
pinMode(led, OUTPUT);
```

As this only needs to be done once, it is placed inside the setup function. Once the pin is set as an output, it will stay an output until we tell it to be something else.

For it to blink, the LED needs to turn on and off repeatedly, so the code for this goes inside loop:

```
digitalWrite(led, HIGH);    // turn the LED on (HIGH is the voltage level)
delay(1000);                // wait for a second
digitalWrite(led, LOW);     // turn the LED off by making the voltage LOW
delay(1000);                // wait for a second
```

The command digitalWrite takes two *parameters* (pieces of data that the function needs to run), which are passed to the function inside parentheses and separated by a comma. The first parameter defines which Arduino pin to write to (in this case, pin 13, as specified by led), and the second parameter gives the value to be written to the pin. A value of HIGH sets the output to 5V, turning the LED on, and a value of LOW sets the pin to 0V, turning the LED off.

The delay function holds the parameter that defines how long the Arduino should continue with its current function. In this case, a value of 1000 delays the program for one second before changing the state of the LED.

CONFIGURING DIGITAL INPUTS

Digital pins can also be set as input pins using the pinMode command. The *Blink* sketch doesn't do this, so here's an example:

```
pinMode(7, INPUT)
```

This pinMode function sets pin 7 as an input. Just as with an output, you'll rarely need to change the mode of a pin, so define input pins in the setup function.

Having set the pin as an input, you can then read the voltage at that pin, as in this example loop function:

```
loop()
{
  if (digitalRead(7) == HIGH)
  {
    digitalWrite(led, LOW)
  }
}
```

Here, the LED will be turned off if the input at pin 7 is read as HIGH at the time it is tested. The Arduino decides whether to turn the LED on with an *if statement,* which starts with the if command. Immediately after the word if is a *condition.* In this case, the condition is (digitalRead(7) == HIGH). The double equal sign (==) tells the machine to compare the two values on either side. In this case, if pin 7 is HIGH, then the block of code surrounded by { and } after the if will run; otherwise it won't. We have already met the code to be run if the condition is true. This is the digitalWrite command to turn the LED on.

NOTE *Lining up the { and } makes it easier to see which } belongs to which {.*

STABILIZING DIGITAL INPUTS WITH PULL-UP RESISTORS

The preceding example code in assumes that the digital input is definitely either high or low. A switch connected to a digital input can only close a connection. You'll typically connect switches in such a way that when flipped, the digital input is connected to GND (0V). While the switch's connection is open, the digital input is said to be *floating.* That means the input isn't electrically connected to anything, but a floating input can still pick up electrical noise from the circuitry around it, causing the voltage on the pin to oscillate between high and low.

This behavior is undesirable because the code could be activated unexpectedly. To prevent input pins from floating, just add a pull-up resistor (Figure C-5). We use just such a resistor in "Project 6: PIR Zombie Detector" on page 72.

When the switch is open (as shown in Figure C-5), the resistor connects the input pin to a voltage source, pulling up the voltage at the input pin to 5V and holding it there. Pressing the button to close the switch overrides the weak pulling up of the input, connecting the digital input to GND instead.

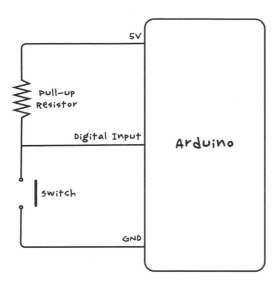

FIGURE C-5: SCHEMATIC FOR USING A PULL-UP
RESISTOR WITH A DIGITAL INPUT

Arduino inputs have built-in pull-up resistors of about 40 kΩ that you can enable as follows:

```
pinMode(switchPin, INPUT_PULLUP);
```

This example shows how you would set the pin mode of a digital input to be used with a switch using the Arduino pull-up resistor: just set the pin mode to INPUT_PULLUP rather than INPUT.

READING ANALOG INPUTS

Analog inputs allow you to measure a voltage between 0V and 5V on any of the A0 to A5 analog input pins on the Arduino. Unlike with digital inputs and outputs, you don't need to include the pinMode command in setup when using an analog input.

You use analogRead to read the value of an analog input, and you supply the name of the pin you want to read as a parameter. Unlike digitalRead, analogRead returns a number rather than just true or false values. The returned number will be between 0 (0V) and 1,023 (5V). To convert the number into an applicable voltage, multiply the value by 5 and then divide it by 1,023, which amounts to dividing it by 204.6.

Here's how you'd read an analog value and convert it in Arduino code:

```
int raw = analogRead(A0);
float volts = raw / 204.6;
```

The variable `raw` is an `int` (whole number) because the reading from an analog input is always a whole number. To scale the raw reading as a decimal number, the variable needs to be a `float` (floating point) type of variable.

WRITING TO ANALOG OUTPUTS

Digital outputs only allow you to turn a component (like an LED) on and off, but analog outputs allow you to control the level of power supplied to a component incrementally. This control allows you to, for example, control the brightness of an LED or the speed of a motor. This is used in "Project 20: Silent Haptic Communication with Arduino" on page 209 to reduce the power to the motor so that it doesn't attract zombies by making too much noise.

Only the pins D3, D5, D6, D9, D10, or D11 are capable of being used as analog outputs. These pins are marked with a little tilde (~) beside the pin number on the Arduino.

To control an analog output, use the command `analogWrite` with a number between 0 and 255 as the parameter, as in the following line:

```
analogWrite(3, 127);
```

A value of 0 is 0V and fully off, while a value of 255 is 5V and fully on. In this example, we set the output of pin D3 to 127, which would be half power.

REPEATING CODE IN CONTROL LOOPS

Control loops (not to be confused with the `loop` function) allow you to repeat an action a set number of times or until some condition changes. There are two commands you can use for looping: `for` and `while`. You would use the `for` command for repeating something a fixed number of times and `while` for repeating something until a condition changes.

The following code makes an LED blink 10 times and then stops:

```
void setup() {
  pinMode(led, OUTPUT);
  for (int i = 0; i < 10; i++)
  {
    digitalWrite(led, HIGH);
    delay(1000);
    digitalWrite(led, LOW);
    delay(1000);
  }
void loop() {
}
```

HOW ANALOG OUTPUTS GENERATE VOLTAGES

IT IS TEMPTING TO THINK OF AN ANALOG OUTPUT AS BEING CAPABLE OF A
VOLTAGE BETWEEN 0V AND 5V, AND IF YOU ATTACH A VOLTMETER BETWEEN
AN ANALOG OUTPUT PIN AND GND, THE VOLTAGE WILL INDEED SEEM TO
TAKE ON VALUES BETWEEN 0V AND 5V AS YOU CHANGE THE PARAMETER
TO analogWrite. IN FACT, THINGS ARE A LITTLE MORE COMPLEX THAN
THAT. THIS KIND OF OUTPUT IS USING pulse width modulation (PWM).
FIGURE C-6 SHOWS WHAT IS REALLY GOING ON.

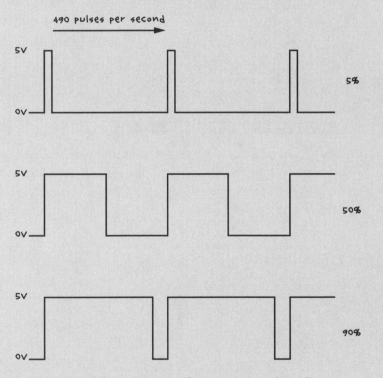

FIGURE C-6: ANALOG OUTPUT'S PULSE WIDTH MODULATION

AN ANALOG OUTPUT PIN GENERATES 490 PULSES PER SECOND WITH VARIED
PULSE WIDTHS. THE LARGER THE PROPORTION OF THE TIME THAT THE PULSE
STAYS HIGH, THE GREATER THE POWER DELIVERED TO THE OUTPUT, AND HENCE
THE BRIGHTER THE LED OR FASTER THE MOTOR.

A VOLTMETER REPORTS THIS AS A CHANGE IN VOLTAGE BECAUSE THE
VOLTMETER CANNOT RESPOND FAST ENOUGH AND THEREFORE DOES A KIND OF
AVERAGING (INTEGRATION).

In this example, we place the blinking code in setup rather than loop, because loop would repeat the blink cycle immediately so the LED would not stop after 10 times.

If you wanted to keep an LED blinking as long as a button connected to a digital input was pressed, you would use a while command:

```
❶ while (digitalRead(9) == LOW)
  {
    digitalWrite(led, HIGH);
    delay(1000);
    digitalWrite(led, LOW);
    delay(1000);
  }
```

This code says that while pin 9 detects that a button is being pressed ❶, the LED should be lit.

SETTING TWO CONDITIONS WITH IF/ELSE

In "Configuring Digital Outputs" on page 251, we used an if command to tell the Arduino IDE to do something if a certain condition was met. You can also use if in conjunction with the else command to instruct the IDE to perform one set of code if the condition is true and a different set of commands if it is false. Here's an example:

```
if (analogRead(A0) > 500)
{
  digitalWrite(led, HIGH);
}
else
{
  digitalWrite(led, LOW);
}
```

This if statement turns the led pin on if an analog reading is greater than 500 or off if the reading is less than or equal to 500.

MAKING LOGICAL COMPARISONS

So far we have used two types of comparison: == (equal to) and > (greater than). Here are some more comparisons you can make:

<=   less than or equal to

>=   greater than or equal to

!=   not equal to

You can also make more complicated comparisons using *logical operators* like && (and) and || (or). For example, to turn an LED on if a reading is between 300 and 400, you could write the following:

```
int reading = analogRead(A0);
if ((reading >= 300) && (reading <= 400))
{
  digitalWrite(led, HIGH);
}
{
  digitalWrite(led, LOW);
}
```

In English, this code might read, "If the reading is greater than or equal to 300 *and* the reading is less than or equal to 400, then turn the LED on." Since we're using the && operator to specify that both conditions must be true, if either condition is not met, the LED remains dark.

GROUPING CODE INTO FUNCTIONS

Functions can be confusing if you're new to programming. Functions are best thought of as ways to group together lines of code and give them a name so that the block of code is easy to use over and over again.

Built-in functions such as `digitalWrite` are more complicated than they first seem. Here is the code for the `digitalWrite` function:

```
void digitalWrite(uint8_t pin, uint8_t val)
{
    uint8_t timer = digitalPinToTimer(pin);
    uint8_t bit = digitalPinToBitMask(pin);
    uint8_t port = digitalPinToPort(pin);
    volatile uint8_t *out;

    if (port == NOT_A_PIN) return;

    // If the pin that support PWM output, we need to turn it off
    // before doing a digital write.
    if (timer != NOT_ON_TIMER) turnOffPWM(timer);

    out = portOutputRegister(port);

    uint8_t oldSREG = SREG;
    cli();

    if (val == LOW) {
        *out &= ~bit;
    } else {
        *out |= bit;
    }
```

```
    SREG = oldSREG;
}
```

Since someone already wrote the `digitalWrite` function, we don't have to worry about what all this code does; we can just be glad that we don't have to type it all out every time we want to change `pin` from `high` to `low`. By giving that big chunk of code a name, we can just call the name to use this code.

You can create your own functions to use as shortcuts for more complicated chunks of code. For example, to create a function that makes an LED blink the number of times you specify as a parameter, with the LED pin also specified as a parameter, you could use the sketch below. This function is named `blink`, and you can call it during startup so that the Arduino *L* LED blinks five times after a reset.

```
❶ const int ledPin = 13;

❷ void setup()
  {
    pinMode(ledPin, OUTPUT);
❸   blink(ledPin, 5);
  }

  void loop() {}

❹ void blink(int pin, int n)
  {
❺   for (int i = 0; i < n; i++)
    {
      digitalWrite(ledPin, HIGH);
      delay(500);
      digitalWrite(ledPin, LOW);
      delay(500);
    }
  }
```

At ❶, we define the pin being used. The `setup` function at ❷ sets `ledPin` as an output and then calls the function `blink` ❸, passing it the relevant pin and the number of times to blink (5). The `loop` function is empty and does nothing, but the Arduino IDE insists that we include it even if it serves no purpose. If you don't include it, you will get an error message when you install the program.

The `blink` function itself begins at ❹ with `void`. `void` indicates that the function does not return any value, so you cannot assign the result of calling that function to a variable, as you might want to do if the function performed some kind of calculation. Then follows the name of the function (`blink`) and the parameters the function takes, enclosed within parentheses and separated by commas. When you define a function, you must specify the type of each of the parameters (for example, whether they are int or float). In this case, both the pin (pin) and the number of times to blink (n) are int values. Lastly, at ❺, we have a for loop that repeats the digitalWrite and delay commands inside it n times.

That's it for the software crash course. If you want to learn more about programming for Arduinos, visit *http://www.arduino.cc/* before everyone at your Internet service provider becomes a zombie.

ASSEMBLING A SCREWSHIELD

Many of the projects in this book use a screwshield that fits over the Arduino sockets and allows you to connect wires to Arduino pins using screw terminals. Not all wires will fit into the normal Arduino sockets, but almost any thickness of wire will fit securely in a screw terminal and won't come loose. There are various screwshields on the market, all with slightly different layouts. In this book, I use the popular model from Adafruit (the proto-screwshield, part number 196), which is provided as a kit that you have to solder together. There are lots of connections to make, but none of them are difficult. The component parts of the proto-screwshield are shown in Figure C-7.

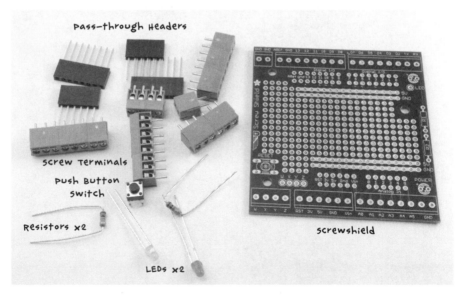

FIGURE C-7: THE PARTS OF ADAFRUIT'S PROTO-SCREWSHIELD

The screw terminals line the edge of the board and Arduino pass-through headers. The screwshield pass-through headers slot through the shield into the PCB. You can plug wires into these as you would in the Arduino Uno, and they have sockets on the top side so you can plug still another shield on top.

Of the two LEDs, one is a power LED that indicates when the board is powered up, and the other is for you to use in your build. You don't have to solder either LED in place if you don't need them. The push button is a reset switch, which can be useful as it's hard to get at the Arduino's reset button when the screwshield is in place. Again, it is by no means essential.

Figure C-8 shows the board being assembled.

FIGURE C-8: ASSEMBLING THE SCREWSHIELD

To assemble the screwshield, follow these steps:

1. Solder the LEDs, resistors, and switch (assuming you want them) in place (Figure C-8a).

2. Put all the screw terminals in place along the outermost edges of the screwshield (Figure C-8b) and flip the board over to solder them on the underside of the PCB. Make sure they are the right way around so that the openings where the wires enter are facing outward, away from the board.

3. Push the pass-through headers through from the top of the board (Figure C-8c) and solder them. Notice that there are two rows of holes on each side of the board where they are able to go; place them in the outer sets of holes. The inner sets are used to wire things up to the pins on the central prototyping area of the board.

If you need a refresher on how to solder to a PCB, review "Soldering Basics" on page 230. With your components in place, make sure your solder joints look sound (also described in "Soldering Basics"). You should be ready to deploy this handy shield in all of your antizombie base defense endeavors and conserve precious solder for devices you intend to last a long time.

FURTHER RESOURCES

There are many great online resources and books that will tell you more about how to use the Arduino in your projects. Here are a few links to get you started:

- I have written a number of books on Arduino, including *Programming Arduino: Getting Started with Sketches* (Tab Books, 2012) and various Arduino project books. You can find a full list of my books at *http://www.simonmonk.org/*.

- Jeremy Blum, the technical editor of this book, has made a great series of introductory videos on the Arduino, which you can find here: *https://www.youtube.com/playlist?list=PLA567CE235D39FA84*.

- Jeremy also has written a great book on Arduino, called *Exploring Arduino* (Wiley, 2013).

- I have written a series of online Arduino lessons, the Adafruit "Learn Arduino" series, which you can find here: *https://learn.adafruit.com/series/learn-arduino/*.

INDEX

The Maker's Guide to the Zombie Apocalypse is set in Schneidler, Billy the Flying Robot, Microbrew, and TheSansMono Condensed. The book was printed and bound by Lake Book Manufacturing in Melrose Park, Illinois. The paper is 60# Husky Opaque Offset, which is certified by the Sustainable Forestry Initiative (SFI).

The book uses a layflat binding, in which the pages are bound together with a cold-set, flexible glue, and the first and last pages of the resulting book block are attached to the cover. The cover is not actually glued to the book's spine, and when open, the book lies flat and the spine doesn't crack.

RESOURCES

Visit *http://nostarch.com/zombies/* for resources, errata, and more information.

More no-nonsense books from **NO STARCH PRESS**

ARDUINO WORKSHOP
A HAND-ON INTRODUCTION WITH 65 PROJECTS
by JOHN BOXALL
MAY 2013, 392 PP., $29.95
ISBN 978-1-59327-448-1

THE SPARKFUN GUIDE TO PROCESSING
CREATE INTERACTIVE ART WITH CODE
by DEREK RUNBERG
AUGUST 2015, 312 PP., $29.95
ISBN 978-1-59327-612-6
full color

THE SPARKFUN GUIDE TO ARDUINO
by DEREK RUNBERG *and* BRIAN HUANG
SPRING 2016, 200 PP., $29.95
ISBN 978-1-59327-652-2
full color

THE MANGA GUIDE TO ELECTRICITY
by KAZUHIRO FUJITAKI, MATSUDA, *and* TREND-PRO CO., LTD.
MARCH 2009, 224 PP., $19.95
ISBN 978-1-59327-197-8

JUNKYARD JAM BAND
DIY MUSICAL INSTRUMENTS AND NOISEMAKERS
by DAVID ERIK NELSON
OCTOBER 2015, 408 PP., $24.95
ISBN 978-1-59327-611-9

LEARN TO PROGRAM WITH MINECRAFT
by CRAIG RICHARDSON
WINTER 2015, 304 PP., $29.95
ISBN 978-1-59327-670-6

PHONE:
800.420.7240 OR
415.863.9900

EMAIL:
SALES@NOSTARCH.COM

WEB:
WWW.NOSTARCH.COM